Contents

Editor's note

An encouraging feature in geographical education in recent years has been the convergence taking place of curriculum thinking and thinking at the academic frontiers of the subject. In both, stress has been laid on the necessity for conceptual approaches and the use of information as a means to an end rather than as an end in itself.

The central purpose of this series is to bear witness to this convergence. In each text the *key ideas* are identified, chapter by chapter. These ideas are in the form of propositions which, with their component concepts and the inter-relations between them, make up the conceptual frameworks of the subject. The key ideas provide criteria for selecting content for the teacher, and in cognitive terms help the student to retain what is important in each unit. Most of the key ideas are linked with assignments, designed to elicit evidence of achievement of basic understanding and ability to apply this understanding in new circumstances through engaging in problem-solving exercises.

While the series is not specifically geared to any particular 'A' level examination syllabus, indeed it is intended for use in geography courses in polytechnics and in colleges of education as well as in the sixth form, it is intended to go some way towards meeting the needs of those students preparing for the more radical advanced geography syllabuses.

It is hoped that the texts contain the academic rigour to stretch the most able of such candidates, but at the same time provide a clear enough exposition of the basic ideas to provide intellectual stimulus and social and/or cultural relevance for those who will not be going on to study geography in higher education. To this end, a larger selection of assignments and readings is provided than perhaps could be used profitably by all students. The teacher is the best person to choose those which most nearly meet his or her students' needs.

W.E. Marsden
University of Liverpool

Preface

"Nature goes her own way, and all that to us seems an exception is really according to order." *Goethe, 1824.*

In writing this book, we have been aware throughout that the physical landscape is an environment of immense diversity and considerable complexity. We have attempted to study the landscape by looking at its individual components. The processes that operate within each system evoke a response which produces a particular landform, and our aim in this book is to examine this interaction between process and landform.

Inevitably, it is impossible to be comprehensive with such a wide field to cover, but we hope we have considered those aspects of geomorphology which will help to unravel many of the problems presented by the working of the British landscape. We have tried to do this as simply as possible, bearing in mind the requirements of sixth form students but, we hope, not dodging the more difficult and less easily comprehensible interrelationships. We have drawn material from both the more familiar descriptive and the newer analytical sources in an attempt to illustrate a variety of methods of approach. Deliberately, whenever possible, we try to relate the concepts we identify to the real world and we have chosen examples to illustrate these from the British Isles where possible. The scale of our study varies widely from micro to macro scale landforms.

Throughout, we encourage students to investigate for themselves the processes and resulting landforms. In the text we have indicated some areas where research, even at a relatively simple level, could provide an interesting and valuable insight into the complexity of geomorphological systems.

Our thanks must go to many friends and colleagues who have contributed ideas, made suggestions and loaned material for inclusion in the book. In particular, we owe much to Bill Marsden whose editorial eye, watchful as ever, improved the structure and systematised the contents of the chapters.

Alan Clowes

Peter Comfort

March 1981

1 Introduction

A. Geomorphology

If we break the word *geomorphology* down into its component parts its meaning quickly becomes apparent: *geo* means earth, *morph* means shape or form, and *ology* means study. A literal translation would give us 'the form of the earth', but the word is specifically used to mean the form of the earth's *surface*. We intend to look not only at form, but also at the *processes* which have been responsible for the formation of that particular form. Geographers in the past have been anxious to catalogue and describe the earth's surface and the features on it. Today we have switched the emphasis to try to answer questions about why particular forms have developed.

The surface of the earth is the meeting point or interface between the gases of the *atmosphere* and the rocks of the *lithosphere* (lithos = rock) (Fig. 1.1a).

Fig. 1.1 The earth–atmosphere system

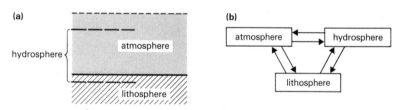

Across this interface pass gases, liquids and solids. Probably the most important of these substances is water. It is present in rocks and in the atmosphere. It forms a zone which lies across the rock–atmosphere boundary and is often referred to as the *hydrosphere*. Each of these areas interacts with the others and we will investigate something of the processes of each (Fig. 1.1b).

B. Scale

Over the 150 million square kilometres of the earth's land surface there is a perceptible variation in shape and process. Ben Nevis and the Norfolk Broads are two very different places to spend a holiday. Yet if we were to see Britain from a satellite, the differences between these two places would be less evident. If we looked at the whole of the northern hemisphere, the differences between Ben Nevis and the Norfolk Broads would be even more difficult to detect, although those between the Alps and the Sahara would be noticeable.

If we were to go the other way and narrow down our field of vision, differences would become more evident. Deposits dumped by glaciers in Norfolk and Scotland might at first sight look rather similar. Closer inspection, however, might reveal that the minerals in the Norfolk deposit could only have originated in Scandinavia, while those on Ben Nevis could only have come from that area. Perception of differences and similarities depends to some extent, therefore, on scale. The matter of the universe is composed of a few types of subatomic particles, mainly electrons, protons and neutrons. It is the way these particles are grouped together to form atoms of the elements, and the way elements combine to form compounds of elements that makes up the variety of the universe. To try to represent the size of the universe and that of the subatomic particle on the same diagram creates a problem. Relative to an atom, 1 mm is a very long way, but on the scale of the universe it is totally insignificant. The problem of the representation of scale is one we will encounter frequently. There are ways around this problem.

Take the diameter of a sand grain, for convenience 1 mm, and the diameter of the earth, approximately 12 000 km. The ratio between these two values is 12 billion to one. If we double each, the ratio remains the same. In Fig. 1.2

Fig. 1.2 The scale of natural phenomena

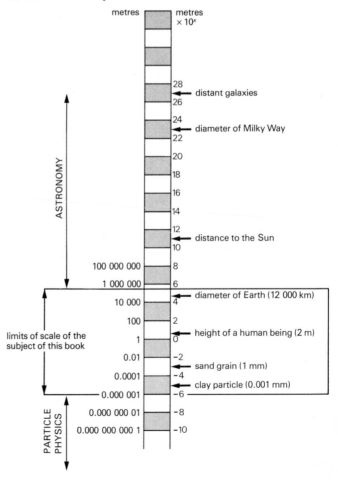

the process of doubling is represented by a uniform increase in the distance along the scale; the distance between 1 mm and 2 mm is exactly the same as that between 12 000 km and 24 000 km. These scales are *logarithmic*. Slide rules are graduated in this way. It is the ratio between values not the values themselves which are important.

The scale of the subject of this book extends from that of the earth down to the smallest movable fragment of rock, a clay particle. These are shown on Fig. 1.2. However, occasionally we need to stretch these boundaries to include the sun, where the energy of most of our systems comes from. We also need to include atomic-sized particles in order to look at the chemistry of some processes. The boundaries are not really very clear. At one end we begin to move into the field of astronomy and at the other to that of physics. The main area of concern is indicated in Fig. 1.2.

A more practical use of logarithmic scales is in the classification of the sizes of particles of rock. Some of these particles are house-sized, occasionally even larger; others are nearly molecular-sized. Fig. 1.3 shows the sizes of particles graphically, from clay, 1/1000 mm across, to cobbles, boulders and even larger fragments. Notice how the scale works by a factor of two so that each jump halves the value or doubles it. The phi (ø) scale uses this system to make the numbers simple. There are twenty steps covering the whole size range.

Fig. 1.3 The size of particles

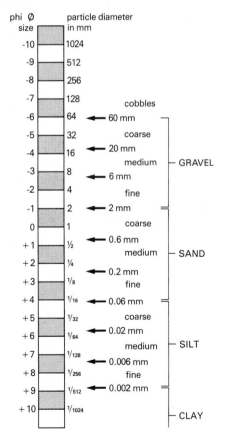

C. Systems

The problem of examining the whole of the natural universe is evidently immense. We have narrowed down the field to look at one compartment, geomorphology. Each chapter of this book represents a further subdivision in which one geomorphological environment of the earth's surface is examined. *Systems analysis* has given us a similar approach, breaking down reality into *elements* and identifying the *linkages* between them. A very simple example of this is a car (Fig. 1.4). Fuel drives the engine which produces changes in the other elements of the system. Ultimately the wheels, in contact with the road, are driven. Most of the links in this system operate in both directions. For instance, a car rolling down hill turns the transmission (axles) round. Particularly interesting is the alternator, which is driven by the engine and produces electricity which is returned to the spark plugs in the engine. In systems terms this is known as *feedback*. The driver receives feedback from the engine when listening to how fast the engine is running and deciding whether or not to change gear.

Fig. 1.4 The car as a system

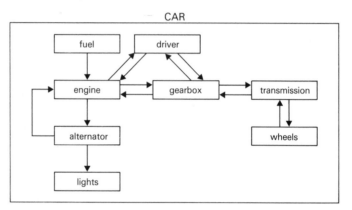

If fuel were fed continuously to the engine, it would go faster and faster and demand more and more fuel. In the end it would break down. This is an example of *positive feedback* which, unless regulated in some way, leads to a breakdown in the system. *Negative feedback* occurs when an increase in the activity of one element of the system induces a decrease in the activity of another. In the case of the engine this is the driver taking pressure off the accelerator, thus restricting the fuel flow. Feedback regulates the system and tends to maintain a *steady state*. Examples of these types of system are detailed in later chapters.

The labels on the boxes in Fig. 1.4 tell us nothing about the complexity of the internal combustion engine. From the diagram we know only that fuel goes in and is changed to another form of energy which is passed on to the gearbox. The engine is regarded as a *black box*. It is not necessary to know how the engine functions in order to drive the car. However, it is possible to look inside the box and examine the engine system. The box becomes transparent as we begin to understand the operation of the systems within it.

D. Systems in Geomorphology

The way in which the atmosphere, lithosphere and hydrosphere interact was outlined at the beginning of this chapter. The only significant input into this system is *energy* in the form of electromagnetic radiation from the sun. Essentially the system is *closed* off from the rest of the universe. On a short time scale, the car is a closed system, which carries its own energy source, petrol; in the longer term it needs injections of fuel if the engine is to function and output energy to the driving wheels. It can therefore be regarded as an open system, with inputs and outputs of both energy and mass in the form of fuel and exhaust.

Fig. 1.5 The geomorphological system

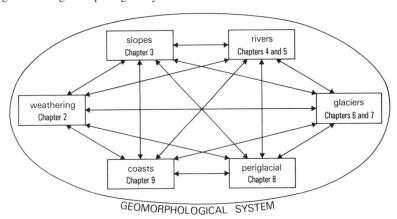

Each chapter of this book looks at an open system which receives inputs, often of rock or water, from other systems (Fig. 1.5). All of these systems are operating within the closed system of the earth–atmosphere. Rivers are a familiar example of an open system. They receive inputs of water mainly from precipitation flowing down slopes, and of sediment derived from the rocks over which the river flows. In populated areas human activity increases sediment and effluent input to the river. The output of the river is often into the ocean, where water, sediment and effluent are deposited. They are generally changed in nature, so the river acts as both a transporting and processing system. Some water evaporates from the river and some may seep into the rocks beneath. There are outputs, therefore, to both the atmosphere and the lithosphere. In Chapters 4 and 5 we take a much closer look at how the river system processes its inputs and outputs.

E. Humans in the System

The surface of the earth is affected by humans to a remarkable extent. Large areas are farmed or grazed, river valleys are often densely populated, roads are built through mountain areas, and coasts are used as ports and harbours. Humans act as an additional element putting a variety of materials into different parts of the geomorphic system. Coal dust from mining has been a significant part of the sediment in rivers on coal fields. The coast has been used as a dumping ground for waste from domestic refuse to toxic chemicals. Not all this input is deliberate; as a result of poor farming practice, soil has been washed from the land surface. Humans have also taken from the system: gravel has been taken from rivers and estuaries for building; slopes have been changed to accommodate roads, and coastlines altered as part of reclamation schemes.

The appearance of humans as another element in the system has often changed the balance achieved by feedback within the natural system. Whole systems, because of their closely interlinked nature, have reacted and changed or have started to change towards a new steady state incorporating humans. Examples of such changes are dealt with in this book. It is important that we understand how these systems operate so that we do not produce changes which may have dramatic and dangerous repercussions. The removal of woodland in one part of a river valley may, for example, cause flooding in another. The Vaiont Dam disaster which is explained on page 53 shows how such consideration of only one part of the system can lead to catastrophic changes in others.

F. The Scientific Method

Many of the ideas in geomorphology originated in the nineteenth century. Pioneer geographers collected information about areas or regions and as the information accumulated they classified it in various ways. This type of activity is characteristic of the early stages of a science. The classifications led to the formulation of *'laws'* and *theories*, many of which have stood the test of time. A number of objections can be made to this method of investigation. It is easy to collect information which is too voluminous to handle and often it is inapplicable since it leads to no valid conclusion.

The *scientific method* was evolved mainly in physics-based science and this provides a more logically acceptable and more concise way to solve problems in science. The starting point is a *hypothesis* which may be accepted or rejected by testing in the real world. Information is collected to evaluate the hypothesis. It is therefore much more particular than the general observations which result from the first method of inquiry.

Fig. 1.6 The scientific method

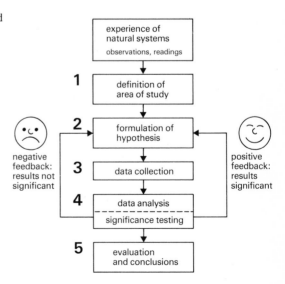

The scientific method is shown diagramatically in Fig. 1.6. There are roughly five formal stages in the process. They are preceded by one similar in form to the first method of inquiry followed by the early geographers, and which consists of personal experiences and informal observations of the earth's surface from which you build up your own mental image of the real world.

1. Defining the problem

Many students find this stage difficult in framing their own investigations. Finding an area of study which interests you is dependent on what you have read, or on your acquaintance with a particular area. For example, you may notice that the beach you visit changes in height. You decide to find out *why*.

2. Erecting the hypothesis

Every investigation needs a starting point. By reading about beaches you will be able to identify important elements in the system which cause the beach to change. Observation may lead you to think that the speed of the wind is the most effective of these. Thus you arrive at the *hypothesis* or theory that 'there is a relationship between beach height and wind speed'.

3. Data collection

To test the hypothesis, information or *data* about the beach height and the wind must be collected. It is impossible to measure the height of the whole beach, so a small part or *sample* of the whole beach (referred to as the *population*) must be used. The selection of a sample should be unbiased. If the narrowest part of the beach were chosen because it would be easier to survey, then the sample would be unrepresentative of the beach as a whole. *Bias* in sampling can be eliminated by using some random method, for example by spinning a coin or throwing a dice. This cuts down the work to manageable proportions. Sometimes someone else will collect the data for you. Wind direction can be obtained from meteorological stations nearby. These are *secondary sources* of information which themselves may have to be sampled.

4. Analysis of data and interpretation

This part of the procedure is determined by your hypothesis. The relationship between beach height and wind speed can be shown graphically as in Fig. 1.7.

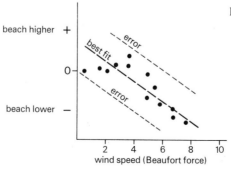

Fig. 1.7 A simple correlation of wind speed and beach changes

The arrangement of the points in an almost straight line shows that there is some correlation between the two variables. In general, the graph shows that as wind speed increases so beach height falls. But not all the points lie in a straight line, implying that there is some degree of vagueness or randomness about the relationship. It is possible that the relationship you have recorded could have occurred by *chance*. This would be more likely if only a small number of points were used in the investigation. If you tossed a coin and recorded two consecutive 'heads' you would not be unduly surprised. If you went on recording heads your suspicions would grow. In general, the larger the sample the more significant are the results. There are statistical tests of significance which could be applied to the beach data. If such a test were to show that your results were significant, then you would be able to verify the initial hypothesis. In Fig. 1.6 this is shown as a positive feedback. If the results were not significant, indicating that they could have occurred by chance, then you need to return to step 2 and revise your views as to the most important variable in determining beach height.

5. Conclusions and evaluation

The conclusion of this investigation may confirm your ideas or those of others about the beach. Perhaps your beach behaves differently than might be expected from the researches of others. The reasons for this are for you to explain and to erect new hypotheses about.

By following the pattern of the scientific method you can be certain that your arguments are well founded and backed up with solid observation.

G. Using this Book

In this chapter we have indicated some of the ideas we think are important and which lie behind the more specific material presented in the rest of this book. In order to help your understanding of the ideas within a chapter, each is followed by three components which are an essential element of the book.

(a) A summary of the *key ideas* contained in the chapter. These are statements of the basic concepts which form the substance of each chapter. Understanding of these indicates an intelligent, thinking, rather than rote learning approach to study.

(b) *Reading* which supplements the material in the chapter. In general the books are easily accessible and fairly easy to understand. Some take you a little further into the subject and may be more useful if you are faced with a particular problem in field work or while conducting your own investigation. These have been marked with an asterisk(*).

(c) *Assignments* which occur both in the body of the text and at the end of each chapter. These are to check your understanding and to help you remember the ideas in the chapter. They vary from short questions, topics for discussion and essay questions to new sets of data for analysis and indications of different ways of thinking about the subject in the chapter.

It is important to emphasise that the summaries of the key ideas, the suggested additional reading and the assignments are not intended as additional extras or time-consuming luxuries. They are intended to be an integral part of this introductory course in geomorphology.

1. (a) *Explain what is meant by the terms morphology and process.*
   (b) *Describe the morphology of a sand castle. Explain the processes which lead to the formation of sand castles.*
2. *Explain briefly the nature of the links shown in Fig. 1.1 between atmosphere, lithosphere and hydrosphere.*
3. *Draw a 'systems diagram' to represent the generation of electricity. The main elements are fuel, dynamo, steam, turbine, water, transformer and national grid. There may be more.*

Key Ideas

A. *Geomorphology*
1. The objective of geomorphology is to explain the form of the earth's surface.
2. Processes operating at the air–rock interface are the mechanisms of change in geomorphology.

B. *Scale*
1. The perception of differences between phenomena is in part a function of scale.
2. Logarithmic scales represent size differences as *ratios*.

C. *Systems*
1. A system is a set of *elements* with *linkages* between them.
2. Systems emphasise the interrelatedness and integration within and between various environments.
3. Many natural systems are self-regulating, having *feedback mechanisms*.

D. *Systems in geomorphology*
1. The earth–atmosphere system is closed and within it a variety of sub-systems, both open and closed, operate at different scales.

E. *Humans in the system*
1. Natural systems must adjust to a new *steady state* when humans appear as a significant element within them.

F. *The scientific method*
1. The scientific method provides the foundation for investigation of natural systems.

Reading

FITZGERALD, B.P., *Developments in geographical method*, Oxford University Press, Chapters 1 and 2, 1974

CHORLEY, R.J., *Physical geography: a systems approach*, Prentice Hall, Chapter 1, 1971

*HARVEY, D., *Explanation in geography*, Edward Arnold, pages 27–43, 1973

2 Rocks and weathering

A. Introduction: Definitions

Many rocks at or near the earth's surface today are found in completely different environments from those in which they originated. They were formed at, or since their formation have been subjected to, high temperatures (e.g. igneous and some metamorphic rocks) or high pressures (e.g. igneous, sedimentary and some metamorphic rocks) in the absence of air and atmospheric water. As a result, surface rocks today slowly undergo changes due to their being in a moist atmosphere under atmospheric pressure and within a temperature range of a few degrees below freezing point to below 100 °C. This process of adjustment is called *weathering* and is the combined action of all processes which cause *disintegration* and *decomposition* of rocks to products that are more in equilibrium with their present environment. The term *regolith* is used to describe weathered rock at the surface extending down to unaltered rock.

It is convenient to subdivide the weathering processes into a *mechanical* (or *physical*) group and a *chemical* group. Physical weathering generally leads to the actual disintegration of rock, which means that the rock is broken apart without any alteration to the minerals which form it. Chemical weathering, however, leads to the decay or decomposition of rock by the alteration or even removal of the rock-forming minerals. It must be emphasised, however, that seldom does physical or chemical weathering operate alone; almost invariably they co-operate with one another to achieve the breakdown of the rocks. No transport of the weathered products is involved except, perhaps, the effects of gravity which may cause them to slip or fall downwards from the site of weathering; such an example is the *screes* along the eastern shore of Wastwater in the Lake District (Plate 2.1).

B. Properties of Rocks Related to Weathering

The number of types of rocks present on the earth's surface is immense, but for various reasons, some are more susceptible than others to the effects of weathering.

The volume of the earth's crust is composed of 95 per cent igneous rocks and only 5 per cent sedimentary and metamorphic rocks. However, these figures are quite different when we consider the actual rocks present and, therefore, exposed to weathering at the surface of the earth in terms of the area they cover.

Plate 2.1 Wastwater screes,
Lake District (Author's
photograph)

1. Physical properties

From Table 2.1 we see that at least 74 per cent of the surface rocks are sedimentary – shale, sandstone and limestone. By reason of their formation, many sedimentary rocks possess *bedding planes*. These bedding planes separated horizontal or nearly horizontal layers of sediment as they were laid down. Thus, after the rocks have been uplifted and exposed in their present day position, bedding planes can function as openings or cracks which the weathering processes, whatever they may be can use.

Table 2.1 Approximate percentage of area of surface rock exposures (Source: Leopold, *Wolman and Miller, Fluvial Processes in Geomorphology*, W.H. Freeman and Company. Copyright © 1964)

Shale	52
Sandstone	15
Granite	15
Limestone	7
Basalt	3
Others	8

Most sedimentary rocks, of which Carboniferous Limestone is an example, also have vertical cracks or *joints* running from one bedding plane to the next. As shown in Plate 2.2, joints, like bedding planes, are places which weathering processes can take advantage of and thus may be regarded as lines of weakness in the rocks. As the joints are enlarged by weathering, so more of the rock surface is exposed to the atmosphere and will result in a faster break-up of the rocks to form what is often described as '*block separation*' (Fig. 2.1).

Joints can also be found in igneous rocks. On cooling from their molten

Plate 2.2 Limestone pavement, Hutton Roof, Lancashire (Author's photograph)

Fig. 2.1 Block separation

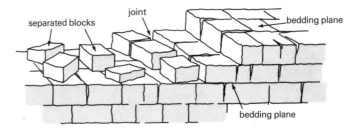

state, some igneous rocks contract and joints are formed. Some granites also show these features, often very unevenly. The hexagonal columns of basalt of the Giant's Causeway on the Antrim Plateau in Northern Ireland (Plate 2.3) are perfectly divided by joints which expose more surfaces to weathering.

Some well-jointed rocks, such as many basalts, decompose readily; the corners and edges are attacked more than the flat surfaces. Thus the rock becomes rounded in all three dimensions. As each 'shell' is weathered away, a fresh 'shell' of rock is exposed and the rock decomposes, becoming more sphere-like (spheroidal) as weathering progresses. This is called *spheroidal weathering* (Plate 2.4).

Cleavage provides other lines of weakness for weathering to attack the rock. It is the direction of easy splitting in metamorphic rocks arising from the parallel alignment of platy minerals such as mica. Plate 2.5 shows how weathering opens up rocks along cleavage planes.

18

Plate 2.3 Basalt columns, Giant's Causeway, Antrim Plateau, Northern Ireland (Author's photograph)

Plate 2.4 Spheroidal weathering of basalt near Giant's Causeway, Antrim Plateau, Northern Ireland (W.E. Marsden)

Fig. 2.2 Spheroidal weathering

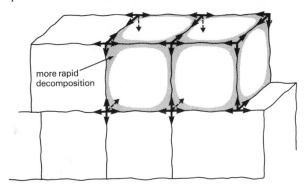

more rapid
decomposition

Plate 2.5 Weathering along cleavage planes, Glyders, North Wales (Author's photograph)

2. Chemical composition

A rock is a collection of various minerals each of which has a different chemical composition. Some minerals are more prone than others to the chemical changes which occur in the context of weathering. For example, limestone, composed of calcium carbonate, is slowly dissolved in rainwater. Many of the silicate minerals found in igneous rocks, such as olivine, pyroxene, amphibole and mica, weather easily (page 24). By far the most resistant of the common rock-forming silicate minerals is quartz. Often quartz is found to be the only remaining mineral after all others have decomposed and this is the reason for quartz sand being such a common deposit in river channels. It is hard, virtually insoluble and, once freed from rocks by weathering, can eventually be transported to the sea to be deposited in a coastal environment.

C. Physical Weathering

As explained in section A, physical weathering involves the mechanical breakdown of rock into smaller particles by the exertion of stresses sufficient to strain and eventually split it. However, no changes occur in the chemical composition of the rocks.

There are several processes by which rocks may be mechanically disintegrated, but the most important fall into three groups. The first two are controlled to a large extent by climate.

1. Crystal growth within a rock

Cracks, joints, bedding planes and cleavage planes are places where water can collect in a rock. If temperatures fall below freezing-point, the ice formed at 0 °C expands by about 9 per cent. Potentially this can create great pressures against the confining walls and in theory a maximum of 2100 kg/cm² is reached at −22 °C. This pressure is about a thousand times that of the average car tyre.

There are very few rocks that could withstand such pressures. In practice, however, such pressures are seldom experienced, but repeated freezing and thawing of water within the rocks does weaken, prise apart and eventually spilt the rocks into angular pieces with sharp corners and edges. This process of rock disintegration is called *frost shattering* or *frost wedging* and occurs commonly in humid climates where day and night temperatures are above and below freezing point respectively on many occasions. The photograph of the Wastwater screes (Plate 2.1) shows a typical locality where frost shattering is operating – a mountainous area above the tree line where there is much surface rock exposure. Frost shattering does not operate where temperatures are always below freezing point, since the ice and snow would never melt and no water could percolate into the rock.

It is easy to oversimplify this process of weathering. Water that has seeped into a rock in some way is never completely sealed in (Fig. 2.3). On freezing,

Fig. 2.3 Frost action

(a)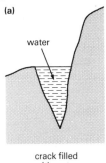

crack filled
with water

(b)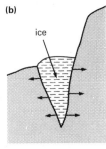

ice extruded above top
of crack on freezing,
some pressure exerted

(c)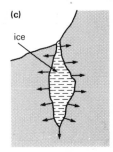

ice more confined,
restricted extrusion above
top of crack on freezing,
more pressure exerted

expansion of the ice could cause it to extrude above the top of the crack, and only part of the expansion would actually be used to build up pressure on the confining walls. Clearly, the more the water is enclosed within the rock, the more effective will be the expansion.

It is not only ice crystals that grow within rocks to cause disintegration but also salt crystals formed by crystallisation from a slightly saline solution. This process operates widely in hot, arid climates, especially in porous rocks, such as sandstone. Here, surface groundwater is lost by evaporation and as a result deeper groundwater is drawn up towards the surface by capillary action. After evaporation, salt crystals may be left behind in pores and cracks in the surface sandstone, growing with additional crystallisation to cause *granular disintegration* of the sandstone – the falling apart of the rock grain by grain.

2. Thermal expansion and contraction

Most solids expand when heated and contract when cooled, and rocks are no exception. It was widely held, therefore, that in areas where large daily temperature ranges were experienced, such as deserts, weathered rock was produced by expansion of the rock's surface layers during the fierce heat of the day and consequent contraction during the clear, cold night. Long continued repetition of this process would weaken the outer 'skin' of the rock which would eventually 'peel' away; thus the rock disintegrates layer by layer, a process known as *exfoliation* (Plate 2.10). However, experiments have proved that the effectiveness of this process has been overrated. In 1936 D.T. Griggs demonstrated under laboratory conditions that a piece of granite subjected to the equivalent of 244 years of daily alternate heating and cooling was unaltered. However, when he used water to cool the specimen rather than let it cool in the air, only two and a half years of weathering were needed to develop cracks and the beginnings of exfoliation. This shows the importance of chemical weathering in the experiment (see section D).

Although this idea of the effectiveness of mechanical weathering fell into disrepute for many years, it is interesting to note that research work reported evidence of this kind of weathering in rocks from the Peruvian desert and, more recently, from the Australian desert that are almost chemically inert.

3. Pressure release and unloading

In the introduction to this chapter it was pointed out that the environment of formation of many rocks was under much greater pressure than the atmospheric pressure exerted on rocks at the surface. Many rocks involved in the crustal uplift and warping associated with mountain building have also been subjected to intense pressures. These rocks were originally buried in a slightly compressed state. As they are exposed by denudation at the earth's surface, and the pressure is released, they expand slightly. This feature is known as *sheeting*, where layers of rock break apart from the rocks below and develop a new form of jointing generally parallel to the ground surface, known as *topographic jointing*. This is only visible in massive, unbedded rocks (Plate 2.6).

Release from the stressed condition can occur when other rocks are removed from above, for example, by fluvial or glacial action. The sheeting thus formed is the result of the so-called 'unloading' of the rock by the erosion agent.

Plate 2.6 Sheeting caused by unloading in an exposed intrusive rock (Eric Kay)

Thus pressure release and unloading contribute to rock disintegration either in itself or, more usually, by opening up the rock for other forms of weathering, both physical and chemical, to operate upon it.

4. Biological weathering

The forcing apart of rocks along bedding planes, joints or even cracks by the ingrowing roots of plants is a fourth, if minor, contribution to physical weathering. Many of these effects have been overemphasised but probably more credit should be given to vegetation in its contribution to chemical weathering, see section D2(b) page 25. Undoubtedly, trees swaying in the wind will weaken the rock and, if blown over, could help to wrench the rock fragments apart.

In addition, physical break-up of rock or fragments may result from the burrowing action of animals or even from the passage of material through earthworms. However, animals seem to have a more important part to play in the way in which they mix soil materials, bringing fresh material to the surface (worm-casts and molehills) where it is exposed to the effects of weathering.

D. Chemical Weathering

The chemical changes and decomposition of rocks at the earth's surface take place in order to create new minerals which are better able to exist in equilibrium with the new conditions of lower pressure, lower temperature, air and water. As with physical weathering, there are several types of chemical weathering, but a major group – *oxidation, hydrolysis* and *hydration* – involves the addition of oxygen and/or water to the chemical structure of the mineral. Both oxygen and water occur of course plentifully in the upper layers of the bedrock and soil. Another group involves the *reaction of natural acids* with the rock minerals to yield salts which may be carried away if they are soluble in water.

1. Oxygen and water reactions

(a) Oxidation

As far as rock weathering is concerned, oxidation is one of the more important reactions. It involves a reaction between metal ions and atmospheric oxygen to form oxides or, if water is included, hydrated oxides. It occurs with water as an intermediary where dissolved gaseous oxygen present in groundwater commonly combines with iron II (ferrous) in mineral compounds to form the more oxidised iron III (ferric). Thus the structure of the original mineral is destroyed and remaining mineral components now are available for further chemical reactions. Oxidised minerals increase in volume, thus the rock itself is weakened by the alteration and becomes more prone to breakdown.

A yellow or reddish-brown discolouration on the surface of many rocks indicates the oxidation of traces of iron that are often present. Oxidation also affects elements other than iron.

(b) Hydrolysis

Hydrolysis is an important chemical reaction between a mineral and water. Igneous rocks, as previously mentioned, are largely composed of silicate minerals and some of these are very susceptible to hydrolysis. The H^+ (hydrogen) ions or OH^- (hydroxyl) ions of water react with the ions of the mineral, thus breaking down the mineral. An example of hydrolysis is the breakdown of potassium feldspar (orthoclase) into the clay mineral, *kaolinite*:

$$2\ KAlSi_3O_8\ +\ 2\ H_2O\ \longrightarrow\ Al_2Si_2O_5(OH)_4\ +\ K_2O\ +\ 4SiO_2$$

| orthoclase | water | kaolinite | soluble potassium oxide | soluble silica or 'silicic acid' |

This feldspar commonly occurs in granites and so, when hydrolysed, granites break down and leave kaolin, a residual solid; like other clay minerals, it is stable under most conditions except tropical climates. Granular disintegration of the granite occurs because kaolin is a soft mineral; it also tends to expand in volume and literally 'burst' the grains apart.

Other clay minerals are formed from hydrolysis of other silicate minerals: *illite* is the residue of hydrolysed sodium and calcium feldspar (plagioclase), and *montmorillonite* is formed from iron-magnesium silicate minerals such as pyroxene (augite) or amphibole (hornblende). Iron released from these minerals on weathering often unites with oxygen and water to form the soft mineral, *limonite*, the cause of the rusty-brown staining of many weathered rock surfaces.

(c) *Hydration*

This form of chemical weathering is the taking up of the whole water molecule (H_2O) into the crystal structure of the mineral. As a result, the altered minerals is enlarged physically, weakening the rock in which the mineral is and perhaps ultimately causing the rock to disintegrate. Some geologists prefer to think of this chemical process as a form of physical weathering related to crystal growth within the rock, see section Cl, page 21.

The conversion of anhydrite to gypsum illustrates the process of hydration:

$$CaSO_4 + \quad H_2O \quad \longrightarrow \quad CaSO_4.H_2O$$
anhydrite   water   gypsum

In this case the uptake of water alters the solubility of the mineral. Gypsum dissolves more rapidly than anhydrite.

Many of the clay minerals described under *Hydrolysis* are hydrated and, in the weathering of minerals such as feldspar, both hydrolysis and hydration may occur together.

2. Acid reactions

(a) *Carbonation*

As rain falls through the atmosphere, the gas carbon dioxide combines with it to form a weak carbonic acid:

$$H_2O \quad + \quad CO_2 \longrightarrow \quad H_2CO_3$$
pure   gas   carbonic acid (rainwater)
water

As an acid, therefore, rainwater reacts with certain minerals. This is very effective in the case of the carbonate minerals, most widespread of which is limestone formed of the mineral calcite (calcium carbonate):

$$CaCO_3 \quad + \quad H_2CO_3 \quad \longrightarrow \quad Ca(HCO_3)_2$$
calcite   carbonic   calcium hydrogen
                 acid   carbonate

Calcium hydrogen carbonate is soluble in water and therefore will be carried away in solution in stream or groundwater.

Carbon dioxide as a gas occurs plentifully in soil, released by the action of decomposition of plant matter. In fact, concentration of carbon dioxide in the 'soil atmosphere' is far higher than that of the normal atmosphere. Thus the acidity of water percolating through the soil increases and carbonation of carbonate minerals, such as limestone, goes on even when not directly exposed at the surface of the ground to rainwater.

(b) *Acids from plant decomposition*

In any soil where decay of plant matter is taking place, several organic acids are formed and are added to the soil water. Such acids may react with the minerals composing the rock, and the products of the reaction will be carried away in solution down into the groundwater, ultimately to add to stream-flow.

3. Tropical weathering

The rate of chemical reactions increases with temperature; for example, an increase of 10 °C doubles or even trebles reaction rates. In the humid tropics, this is shown by the greater depths to which rocks in general have been altered than in temperate latitudes. Rapidly decaying vegetation releases many organic acids, which also speeds up the process of chemical weathering.

In the tropics clay minerals, which are stable in cooler climates, are decomposed further. Weathering removes more minerals from the regolith. For example, silica is more soluble at higher temperatures and may be carried away as silicic acid in solution in groundwater. Other minerals, which are more stable chemically, remain. Very stable hydrated aluminium oxides (*bauxite* when crystallised in soils) or iron oxides (*laterite* when crystallised) may become concentrated as heavy rains leach away other minerals. Sometimes these are indurated (hardened) and may occur within the regolith as *duricrust* or *plinthite* layers. These layers may be exposed by subsequent erosion of the upper regolith. They commonly develop on flat surfaces in tropical lands where, under seasonal rainfall, the water table oscillates, the rocks being saturated for long periods during the wet season. Laterite often sets so hard that after it has been excavated it is used as a building brick (Plate 3.9).

Limestone is another rock which suffers rapid decomposition in tropical areas where carbonation by rainwater and organic acids is accelerated. As a

Plate 2.7 Kweilen tower karst (Jens Bjerre)

result, tropical limestone exposures are often much more affected by weathering than those of temperate landscapes. Extensive surface and underground solution has reduced many limestone areas to much less than their former dimensions.

In South China and North Vietnam this limestone landscape is described as '*cockpit karst*'. The cockpits are depressions in the limestone enlarged by solution of the rock, which is particularly rapid in the monsoon season. The cockpits are separated by cone-shaped hills, which in an extreme form may become limestone towers ('*tower karst*') (Plate 2.7). This landscape occurs on a smaller scale in Jamaica.

ASSIGNMENTS

1. *Describe and explain the contrast in weathering processes operating to form the screes above Wastwater in Plate 2.1 and the steep scree covered slopes in Plate 3.8.*
2. *Make an annotated sketch-diagram of Plate 2.2 (i.e. the rocky area) and explain the processes at work.*
3. *Which subdivisions of both physical and chemical weathering are likely to be more effective in the following circumstances: tropical landscapes, deserts, mountainous areas in Britain, and chalk downland in southern Britain?*

E. Weathering and the 'Debris System'

Having studied the processes whereby weathering occurs, we can now see that most rocks exposed at the earth's surface are progressively either broken down by mechanical means into smaller products or changed by chemical means into different substances. Weathering has a vital part to play in the evolution of the landscape because rocks which are weakened, broken down or altered are all the more susceptible to agents of erosion and transportation, such as running water, the sea, glaciers and ice-sheets or the wind. On a less dramatic scale, but of great importance, weathering prepares rock for the processes of slope formation which is the topic of our next chapter.

We can therefore think of weathering as the 'input' of the 'debris system', as defined in Chapter 1 (page 11). The debris system is an open system and takes account of the way in which weathered products or debris once formed are transported from the site of weathering. As an example, Fig. 2.4 shows the system in a simplified form for a river valley; this should be compared with the system shown in the chapter on Fluvial Landforms, Fig. 5.6, page 126.

Fig. 2.4 The 'debris system'

1. *With reference to the systems diagram (Fig. 2.5), insert the following terms in the relevant spaces to illustrate the weathering processes and their linkages operating on the slopes of a dry river valley in a limestone area:*
    scree formation; solution; block separation; carbonation; joint enlargement.

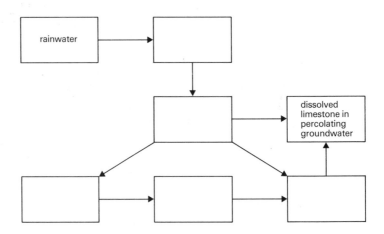

2. *Using a similar method to the above, what differences would there be in the system operating at the top of the Wastwater screes shown in Plate 2.1?*

F. Landforms Produced by Weathering

So far in this chapter we have seen that the most important contribution that weathering makes to landscape evolution is to prepare rock by break-up and alteration for erosion. However, there are some landforms that are almost entirely the result of weathering and these require our attention briefly if our study is to be complete.

1. Tors

Dartmoor, Bodmin Moor and St Austell Moor in the south-west of England are exposures of granitic rocks which are the result of the emplacement of molten rock (magma) long ago in the crustal rocks. Subsequent to their formation, erosion removed the surface rocks to expose these granite *batholiths*. Dartmoor is the most extensive, covering about 550 square kilometres. At various places on each of these upland areas there are masses of granite protruding about 6 to 20 metres through the surface. The exposed granite is known as a *tor*. Plate 2.8 shows a tor in Western Australia where some of the granite is part of the bedrock and the rest is on top of the bedrock.

In 1955 Linton gave one explanation of the formation of Dartmoor tors. Decomposition of the granite below the ground surface took place by weathering to depths of tens or even hundreds of metres, perhaps accelerated by the warmer conditions of the Tertiary era (Pliocene period). The most rapid weath-

Plate 2.8 Granitic tor in Western Australia (I. Kaill)

ering occurred along joints, but where these were widely spaced apart, the rock between would be very little altered. Eventually, as suggested in Fig. 2.6, the surface was lowered by further denudation (erosion and weathering) during the

Fig. 2.6 Stages in development of tors (After Linton)

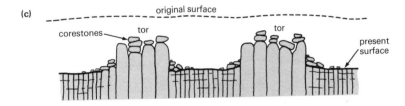

(a) original surface

joints

(b) most rapid sub-surface weathering

slow weathering where joints are widely spaced

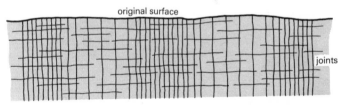

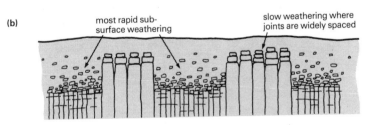

(c) original surface

corestones tor tor present surface

wetter Quaternary era, i.e. periglacial processes (page 214). The weathered granite was removed but the unweathered granite was left as tors standing up above the new lower surface, with *corestones* of unaltered granite resulting from spheroidal weathering forming on top of the tors (page 18). Other explanations of tor formation are given in the asterisked books in the reading list at the end of this chapter.

Tor formation is not restricted to granitic rocks; in the Pennines there are many examples of Millstone Grit tors composed of hard sandstone, around which weathered sandstone has been removed by solifluction (page 57) during a cold phase in the Pleistocene period (Plate 2.9).

Plate 2.9 Millstone Grit tors, South Pennines (Author's photograph)

2. Exfoliation domes

Occasionally single large bodies of massive rock are found where, owing to pressure release and unloading (or sheeting), layers of rock have peeled away leaving what has been called an *exfoliation dome*. Classic examples are found in the Yosemite Valley, California, where these exfoliation domes are dramatic features of the landscape; individual rock sheets of up to 15 m can be found (Plate 2.10).

Plate 2.10 Exfoliation dome, Yosemite, California (Aero Services Corporation)

3. Limestone pavement

On the slopes of Ingleborough in the limestone districts of north-west York-shire, a typical feature is the *limestone pavement* as shown in Plate 2.2. The formation of this feature is the result of chemical weathering along the joints; a limestone surface has been exposed (by previous erosion, perhaps glacial) to reveal a surprisingly regular criss-crossed series of joints. Rainwater percolates down via these joints which are enlarged by the process of carbonation and then solution of the weathered products. Between the *clints* (blocks of limestone) in many of the *grykes* or gaps (enlarged joints), vegetation has begun to establish itself in the thin residue left by the accumulation of impurities from the lime-stone which have not been dissolved away. Release of organic acids further contributes to chemical decomposition and some of the grykes are up to 2 m deep.

It is the opinion of many geologists that such limestone pavements were in the process of formation before exposure at the surface. Under a covering of vegetation, carbonation and solution can be quite effective (page 25); the presence of organic acids is also a factor. This is proved by slight depressions in soil and vegetation, showing that weathering is already attacking the joints between the thinly covered limestone blocks. In other words, limestone pavements may well be much older features, dating back to before the time they were actually exposed to direct rainwater action.

Examination of the surface or sides of the clints reveals solution grooves or *karren* (Plate 2.2) where rainwater has collected in pools and then overflowed into the grykes, dissolving minute amounts of limestone as it goes. Karren several centimetres deep have been recorded; this gives some indication of how slow the rate of chemical weathering is on limestone, taking into account the length of time that such surfaces have been exposed to rainwater.

More impressive in limestone areas, however, is its large-scale removal by solution in flowing water underground to form caves and large passageways.

ASSIGNMENTS

1. *Refer to Plate 2.2. Describe the main features shown and explain the various processes which have been and are at work.*
2. *Consider how rates of erosion affect rates of weathering under different geological conditions. Take account of the various agents of erosion — wind, ice-sheets and glaciers, the sea and rivers — and how their effects may predominate in different environments.*

G. Human Interference

The release of smoke and waste gases into the air from manufacturing industry has altered the proportion of carbon dioxide and sulphur dioxide in the urban atmospheres of many towns, especially in those where there is much industry. This results in more acidic rainwater, carbonic acid and sulphuric acid being more concentrated than usual. Thus buildings made of natural stone are liable to chemical weathering in such areas, and those of limestone are particularly prone to such decomposition.

It is interesting to contrast the effects of atmospheric pollution on buildings in industrial areas with those in non-industrial areas where the atmosphere is less 'toxic'. In these latter areas there is considerably more lichen growth on buildings. Lichens are intolerant of concentrations of sulphur dioxide in the atmosphere; the quantity of lichen growth therefore gives some indication of the degree of atmospheric pollutants present.

Key Ideas

A. *Introduction: definitions*
1. Weathering brings rocks through processes of adjustment to changed circumstances into *equilibrium* with their present environment.
2. Physical weathering generally leads to the *disintegration* of rocks.
3. Chemical weathering generally leads to the *decomposition* of rocks.

B. *Properties of rocks related to weathering*
1. *Bedding planes, joints* and *cleavage* are lines of weakness providing a means by which weathering attacks the rock.
2. The chemical composition of the rock-forming minerals determines their resistance to weathering, especially in relation to chemical weathering.

C. *Physical weathering*
1. Fluctuations of temperature around freezing point cause *frost shattering* if water within a rock is trapped and unable to relieve the build up of pressure by its extrusion.
2. *Granular disintegration* of a rock is the result of a grain by grain break-up caused by porous rock being over stressed by the growth of water or salt crystals.
3. *Exfoliation* shows that outer layers of rock are most susceptible to large daily variations in temperature.
4. *Pressure release* and *unloading* assist other forms of weathering such as *sheeting*.
5. Plants and organisms make small direct contributions to weathering.

D. *Chemical weathering*
1. *Oxidation, hydrolysis* and *hydration* result in alteration of the original minerals by the addition of water and/or oxygen into their chemical structures.
2. Carbonate minerals are particularly attacked by weak acids — carbonic (rainwater) and organic (plant decay).

E. *Weathering and the 'debris system'*
1. The most important role of weathering in most landscapes is to prepare rock for the processes of erosion and slope development.
2. Weathered material provides the *'input'* for many open 'debris' systems operating in different environments.

F. *Landforms produced by weathering*
1. *Tors* indicate that subsurface chemical weathering occurs before rocks are totally exposed at the surface.
2. *Exfoliation* domes show that sheeting alone can be responsible for quite impressive features.
3. Extensive *limestone pavements* composed of *clints* and *grykes* reveal the effectiveness of carbonation and solution.
4. *Karren* prove surface solution even if limestone pavement was formed in part before exposure.

G. *Human interference*
1. Industrialisation has increased the acidity of rainwater, quickening the rates of chemical weathering in such areas; building stone illustrates this.

Additional Activities

1. Old tombstones make an interesting study of rock weathering, since it is possible to work out some idea of the rate at which the rock is being weathered from the date on the tombstone. From a local cemetery, observe how alteration seems to depend on age, rock type and aspect (exposure windwards will probably be wetter). See if sunny surfaces show any differences from those which are always in the shade.
2. Observe how deeply weathering penetrates into pebbles and rocks by breaking them open and measuring the distance from the outside to where unaltered rock is found. Do this in several different areas, examining a

variety of rock types. Try to explain the variations you find with reference to location and type of rock.

3. In an area with which you are familiar (your own local area is ideal), discover the type and composition of rocks that generally make up the hill lands and lowlands. Can you see any relationship between the landscape and the rates of weathering of the different rocks?

4. Conduct your own laboratory experiments with different rocks to test their response to weathering:

   (a) Take samples of sandstone, limestone, shale and mudstone and immerse them in water for a day; allow them to drain for the same length of time. Repeat this for several weeks and see if any breakdown is occurring. Describe and explain what is happening.

   (b) Obtain a piece of porous rock, for example chalk, sandstone or oolitic limestone. Soak it in water for two to three hours, then put it in a freezer for about three hours. Immerse it again when it has thawed out and then refreeze it. Repeat this process several times. If this represents what happens daily to rocks in cold areas, how long would it take for frost shattering to effect rock disintegration under natural conditions?

Reading

OLLIER, C., *Weathering and landforms*, Macmillan, 1974

BLOOM, A.L., *The surface of the earth*, Prentice Hall, pages 16–39, 1973

STRAHLER, A.N., *Physical geography*, Wiley, pages 395–402, 1975

SAWYER, K.E., *Landscape studies*, Edward Arnold, pages 1–4, 120–127, 1975

SPARKS, B., *Geomorphology*, Longman, pages 22–48, 1972

*SPARKS, B., *Rocks and relief*, Longman, 1973

*SMALL, R.J., *The Study of landforms*, Cambridge University Press, pages 15–27, 1978

*PITTY, A.F., *Introduction to geomorphology*, Methuen, pages 172–173, 184–199, 1971

*RICE, R.J., *Fundamentals of geomorphology*, Longman, pages 115–138, 1977

3 Slopes

A. Introduction

Contours are so much a part of topographical maps that we frequently take them for granted. They denote a change of altitude of the land surface and hence indicate the presence of a slope. The variation in the spacing of contours expresses changes in the angle of slope. There are few areas where the land surface is perfectly horizontal over any significantly large area and contours are absent. Similarly, there are very few areas where the land surface is very steep. The steepness of a slope is a term which is understood through our experience of movement. The most 'energetic' way to travel up a slope is to move up the shortest dimension of the slope, i.e. at right angles to the contours. All other routes are less steep and less 'energetic' since work is spread over a greater distance.

B. Slope Form

The steepness of the slope is measured as the *angle* between the horizontal and the *slope* (Fig. 3.1), measured down the line of the steepest slope. At a point

Fig. 3.1 Basic dimensions on slopes

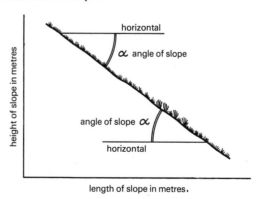

length of slope in metres.

on any hillside there is only one such direction. The *length of the slope* is the *horizontal* distance over which the slope extends. The *height of the slope* is the *vertical* distance between the top and bottom of the slope. The cross section of a slope is the *profile*. The extent of a slope is usually determined by points where the direction of slope changes, at the top of a hill or in the bottom of a valley. It is unusual to find slopes which are completely straight from top to bottom. The angle of slope varies along the profile. Parts of the profile which have a constant angle of slope are called *segments*. We may think of the slope as composed of a large number of shorter segments which may form a curved part of the profile. These curved sections are termed *elements* and may be either *concave* inwards or *convex* outwards from the hillside (Fig. 3.2). In slope

Fig. 3.2 Slope segments and elements

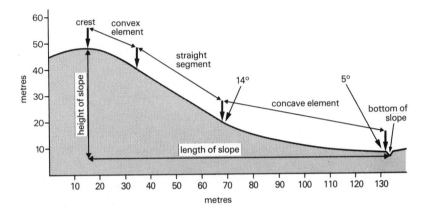

elements the change in the angle of slope from segment to segment, when expressed as a rate per 100 m along the profile, is the *curvature*. In Fig. 3.2 the curvature of the concave elements is –15°/100 m. The slope of the top segment of the elements is 14° and that of the bottom segment 5°, a difference of 9° in a distance of 60 m.

$$\text{Curvature} = \frac{\text{angle of base of slope} - \text{angle of top of slope}}{\text{length of slope}} \times 100$$

$$\frac{5 - 14}{60} \times 100 = -15°/100 \text{ m}$$

By convention concavities are denoted by a negative value and convexities by a positive value. The simple slope shape shown in Fig. 3.2 is a form shown by many slopes but it is by no means the only slope shape. Many natural slopes are much more complex. A number of segments and elements are shown in the profile in Plate 3.1 and Fig. 3.3. There is a short convexity (X) at the summit, followed by a long straight segment and then a series of concavities (V) and alternating convexities linked by steep sections of bare rock (F). Steep rocky faces often occur in profiles and are termed *free faces*. They are the source areas of much of the material which builds up the lower slopes.

1. Slope angles

The area shown in Plate 3.1 is in the high mountains of Norway. It might be expected that such an area would abound in very steep slopes. In fact vertical free faces occupy a very small part of the slope profiles. The angle of slope of

Plate 3.1 Mountain slope profiles in Visdal, Norway (Author's photograph)

Fig. 3.3 A sketch of slope profiles in Visdal, Norway

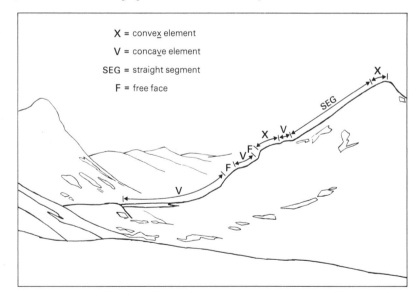

X = conve<u>x</u> element
V = conca<u>v</u>e element
SEG = straight segment
F = free face

the long segment of the profile approaches 35°. Slopes of 45° or more are usually free faces and the province of the mountaineer. No appreciable thickness of *regolith* (page 16) can accumulate and such slopes are recognised as cliffs.

37

Table 3.1 indicates the meaning of slope angles in everyday terms. Most cars can climb a gradient of 1 in 3, a slope angle of 18.5°, which falls into the 'moderately steep' category. The steepest railway gradients are about 1 in 40, approximately a 5° slope.

Table 3.1 Slope angles and their meanings (After Young)

Slope angle (gradient)	Description	Nature
Greater than 45° (1:1)	Cliffs	Usually free faces common only in mountain areas.
30–45° (1:1.7)	Very steep	The steepest slopes on which debris lies. Usually screes and other slopes showing signs of rapid movement.
18–30° (1:3)	Steep	Generally too steep for agriculture. Terracing in the tropics.
10–18° (1:6)	Moderately steep	The upper limit of mechanised cultivation.
5–10° (1:12)	Moderate	Soil erosion in dry areas.
2–5° (1:29)	Gentle	Often depositional areas, flood plains.
Less than 2°	Level	Often depositional areas, flood plains.

2. Slope profiles

If we consider a slope to be made up of a large number of short, straight segments of uniform length, say 10 m, then it will be observed that some of the angles of slope occur more frequently than others. In Fig. 3.3, the long segment below the summit convexity would produce a large number of 10 m lengths with an angle of slope around 35°. On a graph of slope angle against slope length (Fig. 3.4) this would form a peak or *mode* in the distribution. A mode with such a high value is unusual except in a mountainous area. A large number of profiles surveyed by Young in the Derwent basin in the Peak District showed a major mode at 4.5° with a secondary mode at 9° on shales and sandstones of the Millstone Grit Series.

Profiles in central Wales on Lower Paleozoic shales showed no strongly developed mode and there was a fairly uniform distribution of slope angles from 0 to 30° and then a tail off in frequency to angles of a little over 40°.

Modes in the distribution of slope angles are called *characteristic angles*. On the same rock in different areas the characteristic angles may be totally different. The factors which control the characteristic angle are complex and variable (page 48).

Fig. 3.4 Slope histogram

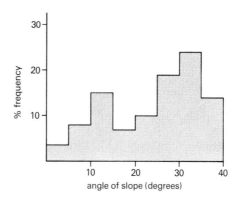

3. Methods of surveying slopes

Surveying slopes is a relatively simple process which yields a great deal of useful and relevant information. There are two principles involved, one which measures the *angle of slope* directly, and the other which measures the change of vertical height on the slope by *levelling*.

(*a*) *Measuring slope angles*

This method involves the measurement of two variables:
 (i) the length of the slope segment,
(ii) the angle of the slope to the horizontal.
There are a number of ways of obtaining these measurements.

(*i*) *Tape and abney*   Using a tape to measure the length of a straight slope segment is an easy operation. Deciding where the segment starts and ends is much more difficult. In geomorphology we are normally concerned with the larger scale variation of the land surface; the small irregularities shown in Fig. 3.5 can be ignored. In some cases the *microrelief* features are very important. If we arbitrarily call slope segments of less than 1 m microrelief, then we have a lower limit to the length of slope segment we wish to study. In Fig. 3.3 a single slope segment is far longer than the usual length of tapes. It is much easier to use a short, *fixed* length of segment of 10 m or 5 m and survey the whole slope in units of that length. If the slope is long and gentle, it is faster and more accurate to use the longer length, whereas on a steep, irregular slope more detail can be obtained using the 5 m length or even shorter.

Having defined the ends of the segment, it remains to measure the slope angle over this length. In Fig. 3.5a the two ranging poles (or broom handles) are the same length and vertical so that the line joining the two tops is parallel to the slope. Measuring the angle between this line and the horizontal is the same as measuring the angle between the slope and horizontal.

An abney level is a small sighting instrument which measures the angle between the line of sight and the horizontal. By placing it on the top of one pole and sighting to the top of the other, it gives the angle of slope α. There are two possible lines of sight, one up the slope and one down the slope. They should

Fig. 3.5 The tape and abney method of measuring slope profiles

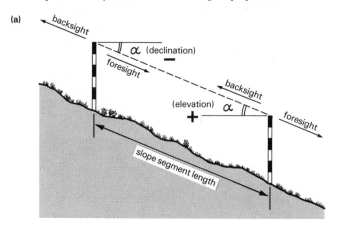

(a)

(b)

Slope location	PIKE HILL		Geology	Permian Sandstones and Marls.	
GR top of slope	764183				
Orientation of profile	174° (GRID)				
Segment no.	Length (m)	Foresight	Backsight	Corrected slope angle (sign as foresight)	Comments
1	9.4	+ 2.5	− 2·5	+ 2.5	CRESTAL SEGMENT
2	10	−1	+ 0.5	− 0.75	
3	8	− 4	+ 4	− 4	
4	11	− 11	+ 11	− 11	Bare rock exposed.

be the same, and reading both provides a useful check. It is easier to survey slopes by this method travelling downhill; the downhill sighting is the *foresight* and the up hill is the *backsight* (Fig. 3.5a).

When the first reading is complete, the pole at the top of the slope moves down the profile to the fixed distance below the lower pole; the process is

Plate 3.2 A small hand-held level sight (Griffin & George Ltd.)

40

repeated until the bottom of the slope is reached. A sample recording sheet is shown in Fig. 3.5b. *Declinations*, angles below the horizontal, are recorded as minus values and *elevations* as positive values. There is space on the right of the sheet for notes to be made about the general nature of the slope.

It is usual to measure slope angles with an accuracy of half a degree and an alternative to an abney is easily made which will measure angles with such accuracy. Figure 3.6b shows a simple clinometer made from a large protractor and a weighted plumb-bob. The sight along the straight edge of the protractor is made from a plastic drinking straw. In use, the object is sighted, the plumb-bob is allowed to settle and is then trapped against the protractor with the finger. The angle of elevation or declination is read as the angle between 90° and the plumb-bob line. By repeating readings, this simple instrument can be made to give quite accurate results.

(ii) Pantometer   The principle of the pantometer is the same as that of the tape and abney method. The pantometer is shown in Fig. 3.6a. Its sides form a parallelogram so that when the short rails are vertical the longer rails are parallel to the slope. The slope angle is read on the protractor fixed to the short vertical rail.

Fig. 3.6 Simple slope measurement instruments: a) a Pitty pantometer and   b) a home-made clinometer

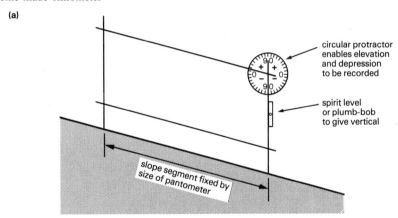

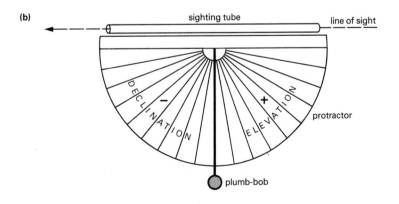

The pantometer was devised by Pitty who initially made the long rail 5 ft long. It is probably better to extend this length to 2 m, although it does make handling and transport a little more difficult.

In use the pantometer can be handled by one person who 'walks' it down the slope moving one leg at a time. The results can be recorded in the same way as with the tape and abney.

Fig. 3.7 Methods of slope levelling: a) levelling principles, b) a sample recording sheet, c) the ruler method

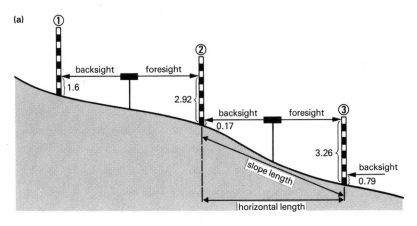

(a)

(b)

Staff position	Length (slope)	Backsight (B)	Foresight (F)	Rise/fall (B-F)	Height	Notes
1	12.4m	1.6			100m	from Bench Mark
2	13.2	0.17	2.92	–1.32	98.68	
3	10.4	0.79	3.26	–3.09	95.59	
4						

Location PIKE HILL
GR top of slope 784191
Orientation of profile 192 (grid)

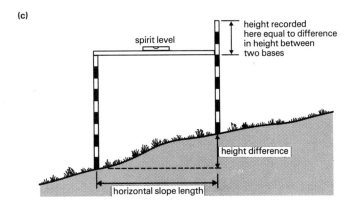

(c)

(b) Levelling

This is the classic way of surveying slopes. It measures *vertical* height changes and *horizontal* distances. The results are easily plotted and enable accurate profiles to be drawn, but require processing to enable angular data to be obtained.

The principle of levelling is illustrated in Fig. 3.7a. The height difference between two *staff positions* is measured using a level line of sight. The surveyor's staff is graduated in 1/100 m and so it is possible to obtain very precise results. In Fig. 3.7a a backsight reading is taken onto the staff, the staff is then moved to the next position down the slope and the foresight reading taken. The difference in height between the two staff positions is calculated by taking the foresight from the backsight; this is entered in the Rise/fall column on the recording sheet (Fig. 3.7b). The level is then moved to position 2 and the backsight is read, then the staff is moved and the foresight is read and recorded. This procedure is continued to the bottom of the slope when the height of each staff position is computed.

Levelling is most accurate if the distance between the level and the staff is kept short and constant. This equalises errors which may arise due to inaccuracy in the level. It also gives a more detailed profile.

There are a number of instruments which can be used to give a level line of sight. Precise levels, such as a dumpy-level, are very accurate but rather expensive. Much cheaper levels, such as that illustrated in Plate 3.2 or water levels made from clear plastic tubing half filled with coloured water, can give reasonably accurate results if used carefully. Some dumpy-levels are fitted with optical devices to enable the horizontal distance to be measured directly. With the other instruments a tape must be used to measure the slope distance.

When the profile to be surveyed is small or a large amount of detail is required, it is possible to simplify the levelling method still further. Using three 1 m rules and a small spirit level, start at the bottom of the slope and support two of the rulers vertically 1 m apart (Fig. 3.7c). The zero ends of the measuring scales should be at the top of each ruler. Place the third ruler on top of the lower of the two verticals and touching the other. If the spirit level is glued to the third ruler it is easy to adjust it to a level position. The difference in height between the bottoms of the two rulers is given by the reading on the higher of the two where it is crossed by the horizontal. Levelling then proceeds in the normal way, 'leapfrogging' the rulers up the slope.

4. Slopes in plan

This chapter began by mentioning contours. They are rarely perfectly straight for any distance, neither are the slopes which they represent. Slopes form spurs, valleys, cirques, hills and ridges which are all associations of slopes arranged in particular patterns. Much of this book is concerned with how such patterns arise. Irrespective of origin, there are three types of slope plan analogous to the three main types of slope profile units.

(a) Types of slope in plan

(i) *Straight slopes*, in which lines drawn at right angles to the contours, *orthogonals*, are straight and parallel. These lines are the line of steepest slope

Fig. 3.8 Slopes in plan (After Young)

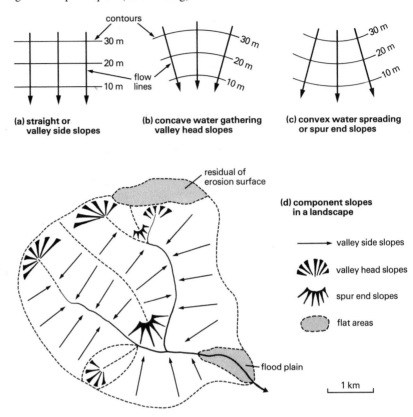

(a) straight or valley side slopes

(b) concave water gathering valley head slopes

(c) convex water spreading or spur end slopes

(d) component slopes in a landscape

→ valley side slopes

valley head slopes

spur end slopes

flat areas

residual of erosion surface

flood plain

1 km

and the line of the slope profile (Fig. 3.8a). They are the *valley side* slopes shown in Fig. 3.8d.

(ii) *Concave plan slopes*, in which the contours form a concave shape into the hillside. The orthogonals to the contours on these slopes converge towards the bottom of the slope (Fig. 3.8b). They form the *valley head* slopes in Fig. 3.8d.

(iii) *Convex plan slopes*, in which the contours are convex out from the hill and the orthogonals spread out from the top of the slope (Fig. 3.8c) and form the *spur end* slopes in Fig. 3.8d.

If we consider the orthogonals or profile lines as flow lines along which water would flow under the influence of gravity, then the concave plan slopes become *water gathering slopes* typically found at *valley heads*, and the convex slopes become *water spreading slopes* typical of *spur ends*. Straight slopes might then be called *valley side slopes*. These properties of slopes have influenced and continue to influence the development of the slope.

(b) Slope maps

(i) A *contour map* is a type of slope map which, by presenting information about the height of the land surface, enables us, with experience, to interpret

44

Fig. 3.9 Morphological maps: a) symbols and b) a map based on an area in south Dyfed

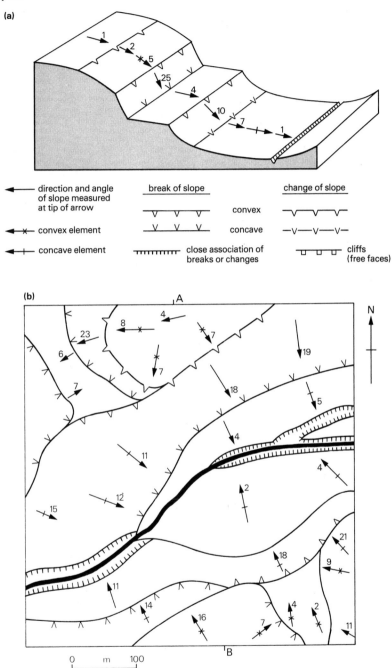

information about the profile and shape of the slope. Closer contours indicate steeper slope angles, while widely spaced contours indicate areas of gentle slopes. It is possible to delimit areas on the map where the contours occur at

45

approximately uniform densities. If the average spacing of the contours is calculated within these areas, the slope angle is given by the formula:

$$\frac{\text{Average contour interval (m)}}{\text{Average spacing between contours (m)}} = \text{Tangent of slope angle.}$$

This technique is particularly useful in analysing changes over large areas from existing maps. Extensive fieldwork may not be possible but the map will enable us to pick out major changes of slopes, such as that from a plateau to a coastal plain.

(ii) *Morphological maps* trace the junctions between different segments and elements identified in the profile over the area of the map. For example, the junction between a concave element and a steep segment such as the base of a scarp is marked on the map by a linear symbol in Fig. 3.9b.

Breaks of slope are points where the transition from one type of slope to the next are rapid, and *changes of slope* are where the transition is more gradual. The symbols used generally are shown in Fig. 3.9a, and a sample of an area in south Pembroke, Wales, mapped by these methods is illustrated. Morphological maps are difficult to interpret and require the mapmaker to walk over all the area to be mapped. They are only a practical proposition when using maps of scales around 1:10 000 or larger, and in the mapping of small relatively distinct features such as glacial moraines (page 189) or beach features (page 249).

ASSIGNMENTS

1. (a) *Using the data in Table 3.2, construct a profile of the slope with the same vertical and horizontal scales. The easiest way to plot profile data is to start at the top of the slope and first mark on the angle of slope. Then scribe off the length of the segment with compasses. Repeat the process to the bottom of the slope.*

    (b) *Identify visually linear segments, and concave and convex elements in the profile and label them in the same way as those in Fig. 3.3.*

    (c) *Label the segment of maximum slope.*

    (d) *Construct a slope frequency graph (slope angle against total slope length of that angle) as described on page 38. Identify characteristic angles in this distribution. How might these be related to the geology of the area?*

    (e) *A third way of representing slopes is to graph the data in the way shown in Fig. 3.10a. This is called a profile graph. Use Table 3.2 to construct a profile graph. Which type of slope unit does this enable you to identify?*

2. *What is profile curvature? Calculate the curvature of the elements you have identified in 1(b).*

3. (a) *In Fig. 3.9b a small area of Pembroke is shown on a morphological map. Use the symbols in Fig. 3.9a to draw the sketch profile along the line AB.*

    (b) *Describe the slope profile constructed in (a). Use the information in Table 3.1 to assist your description. Compare this profile with that of the slope in the valley of the River Dane.*

Table 3.2 A slope profile surveyed in the valley of the River Dane, Cheshire, on a sandstone/shale sequence covered with thin boulder-clay

Number of segment from top of profile*	Angle of slope	Number of segment from top of profile*	Angle of slope
1	+ 1.5	27	−26
2	− 2	28	−25
3	− 1.5	29	−21
4	− 1.5	30	−17
5	− 3	31	−10
6	− 3.5	32	−11
7	− 4	33	−14
8	− 5.5	34	−11
9	− 6	35	−10
10	− 6.5	36	− 9
11	− 7.5	37	− 6.5
12	− 9	38	− 7
13	−11	39	− 5.5
14	−13	40	− 3
15	−15	41	− 2
16	−18	42	− 2.5
17	−21	43	− 1.5
18	−24	44	− 7
19	−25	45	− 9
20	−26	46	−10
21	−27	47	−10
22	−24	48	− 6
23	−24	49	− 3
24	−25	50	− 2
25	−25	51	− 2
26	−27	52	− 1.5

+ values are evalations
− values are declinations
* The length of all segments is 5 m.

4. (a) Differentiate between concave and convex slope profiles, and between water spreading and water gathering plan shapes.
    (b) There are nine possible combinations of the three profile shapes and three plan shapes (page 43). Complete the matrix in Fig. 3.10b. In each cell draw an annotated block sketch or contour map to show the combinations of plan and profile, e.g. a concave water gathering slope. This slope classification can accommodate almost any land surface.

Fig. 3.10 (a) A slope profile graph and (b) combinations of plan–profile

(a)

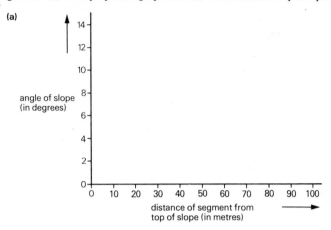

angle of slope (in degrees)

distance of segment from top of slope (in metres)

(b)

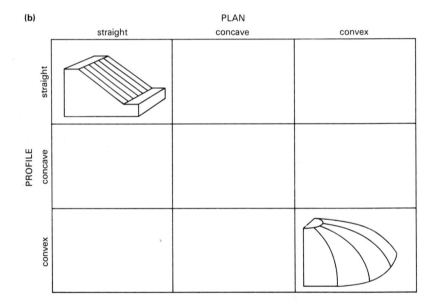

PLAN

straight concave convex

PROFILE

straight

concave

convex

C. Slope Processes and Landforms

Gravity is the principal force affecting slope processes. We may consider it as acting vertically downwards from the centre of mass of any particle (Fig. 3.11). It has two effects:

(i) to *slide* the particle down the slope,
(ii) to *stick* the particle to the slope.

The particle shown in Fig. 3.11 rests on a surface which is slowly tilted until the slope angle is just steep enough to start it sliding down the slope. At this point the frictional forces which act up the slope exactly balance the 'slide' forces which try to pull the particle down the slope. This point is the *angle of friction*. The angle of slope is the angle of friction of the particle. If the slope

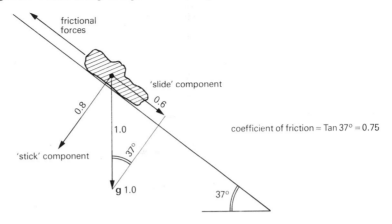

Fig. 3.11 Forces acting on a particle on a slope

were any steeper, the particle would begin to slide. The stick component is just sufficient to hold the material in position on the slope. The tangent of this angle i.e. the ratio between the slide and stick components, is the *coefficient of friction*. Angles of slope greater than 45° which would have a coefficient of greater than 1.0 are rare in natural conditions on regolith slopes.

Regolith moves down slopes as low as 3° in special circumstances (page 216). The coefficient in such a case would be 0.052, the tangent of the angle of the slope. Low coefficients like this are the result of the presence of water in the slope regolith material. It lubricates the contacts between the particles when present in quite small quantities and, more importantly, it fills the spaces (*pores*) between the particles and forces them apart when it is under pressure. This is called *pore pressure* and it becomes increasingly important in slopes of lower angles.

Gravity, slope angle and water contents are the most important factors in determining the nature of movement on slopes. The interplay between them and a larger number of other factors produce such a large variety of slope processes that detailed understanding of them is far from complete.

1. Falls

Falls usually occur on the steepest slopes of 70–90°, where the angle of friction is greatly exceeded. Only hard rocks are capable of maintaining such angles for any length of time, and we have seen that such slopes occur relatively rarely. Falls occur when the internal strength of the rock is overcome. Gravity alone is incapable of causing the rock to break or *fail* and other forces are needed to assist. Ice is often the agent involved (page 21), making this type of movement most common in cold mountainous areas. Thermal disintegration and exfoliation (page 22) produce falls in the hotter areas of the world.

When the particle becomes detached it moves down the slope by a combination of falling, rolling and bouncing until it reaches a point where the slope angle is low enough to allow the particle to come to rest. In terms of our initial analysis, this is where the slope is less than the angle of friction for the particle and the prevailing conditions on the slope.

(a) Screes

Rock fragments detached by the methods outlined above are usually pebble-sized or larger. The typical landforms they produce are *screes* (Plate 2.1). These piles of unconsolidated rock fragments have surface slopes of between 34 and 40°. Most typically the slopes range between 36 and 38° on actively forming screes. Where the screes are inactive, the slopes are usually less steep and there is also a cover of vegetation.

The activity of a scree is dependent on a number of factors related to the rock type, the morphology of the cliff and the climate.

In Chapter 2 it was indicated that the spacing of planes of weakness (joints, bedding and cleavage) determine the size and shape of the weathered particles (page 17). Heavily fissured rocks tend to weather faster than massive rocks. Slates in Snowdonia tend to have fewer free faces than the volcanic rocks. At Tremadoc, North Wales, an igneous rock, dolerite, forms free faces, while the mudstones beneath are largely scree covered because they have undermined the dolerite and caused it to collapse.

Fig. 3.12 Screes: a) development of curved bedrock profiles and   b) the effect of cliff height on scree form

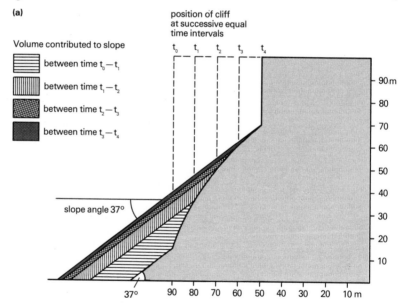

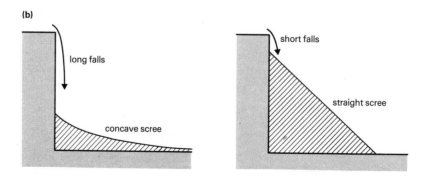

The height of the free face in a cliff of uniform lithology is directly proportional to the rate of supply of material to the scree below it. The higher the cliff the more rapidly the scree accumulates. As the scree builds up it begins to cover the lower parts of the cliff preventing further weathering and fall (Fig. 3.12). The upper part of the cliff retreats at a constant rate — t_0-t_1, t_1-t_2, t_2-t_3, etc. — while the lower parts of the cliff are progressively slowed to zero rate. The scree covers the solid rock, which develops a curved profile. The top of this profile eventually attains the same angle of slope as the scree, 37° in Fig. 3.12a, and forms a straight slope across which a thin layer of debris travels. This appears to be fairly common on many screes, the bedrock often showing through at places.

Cliffs which have long falls onto screes tend to form slightly concave profiles with lower angles, possibly as a result of the large amount of kinetic energy in the rockfall which enables a greater number of rocks to reach the lower parts of the slope (Fig. 3.12b and Plate 3.1).

Cliffs in plan are rarely straight. They form concave shapes (page 44) down which rockfall is channelled. These are called *avalanche couloirs* in high mountain areas, and form large *scree cones* at their lower end (Plate 3.3).

The sorting of debris in screes is very common. The largest boulders are more frequently found towards the base of the scree, because of their greater

Plate 3.3 Scree cone at Cautley, North Pennines (Author's photograph)

momentum and the action of smaller particles as 'bearings' over which the boulders may slide (Fig. 3.13). A fringe of large boulders at the foot of a scree is called *boulder apron*.

It is very difficult to tell visually how active a scree is. Many are not vegetated because the large 'pores' between the fragments allow water to drain away very rapidly. Screes are very dry and most plants would suffer from water shortage. Rapid drainage of water into the slope means that the fine particles are rapidly washed deep into the scree and it is a long time before the water retaining capacity of the scree is increased. Lichens, an association of an alga and a fungus, can grow quite well on the rock particles in a scree and they provide a good indication of how frequently a boulder is disturbed. Larger lichens indicate longer undisturbed periods.

Fig. 3.13 Scree sorting

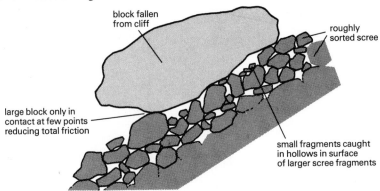

(b) *Measuring movement on screes*

The rate of rock fall from a cliff can be measured by placing boxes or polythene squares at the top of the scree to catch the falling material. It must be remembered that the rate will vary greatly with the siting of the equipment. Random spacing will give a much more representative idea of the rate of rockfall, although in Britain at least a year should be allowed to elapse before any results are assessed.

The rate of movement on the screes can be assessed by using boulders painted with marine paint. If these are laid in lines parallel to the contours, or lines are painted across the boulders already on the surface of the scree, distortion of the line will indicate where movements are greatest. In order to check total movement, the line should be referenced to securely marked positions on bedrock. These can be painted or chiselled into the rock.

Movements are usually seasonal, relating to cold periods when snow or ice may add to the mass of the material on the slope, or push it down slope by expansion. Rates of movement are highly variable even on one slope. Adjacent boulders may show movements of zero and tens of metres. Cliff recession calculated from studies of rockfall vary from 2–13 mm in a semi-arid climate to 0.01–0.1 mm in a polar climate.

2. Slides

Movements of rock and regolith which occur with relatively little disruption or internal deformation in the mobile mass are collectively termed *mass movements*. The amount of deformation varies greatly and the category 'slides' ranges from falls through to highly fluid mudflows (page 57).

Slides affect both hard rocks and unconsolidated regolith material. The moving mass slides down an inclined plane more or less intact until it reaches the bottom of the plane where impact usually breaks it up (Fig. 3.14). In Glen Etive, Argyll, a massive rockslide occurred along sheeting joints (page 22) in a granite and exposed the slide plane over a vertical distance of almost 350 m.

Fig. 3.14 Rapid mass movements

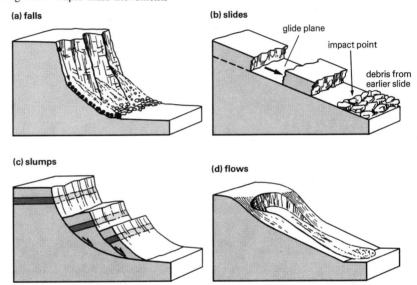

The mass of regolith material and solid granite formed a series of mounds at the base of the new cliff which slopes at 50–55°. The slide was probably triggered off when rainfall added to the mass of the material above the slide plane, producing a greater stress than the layer of granite could withstand. Failure probably occurred along a slight weakness, a joint or mineral vein, at right angles to the slide plane.

A massive rockslide occurred in Northern Italy in a tributary valley of the River Piave, north of Venice, with disastrous results. The Vaiont River had cut a deep gorge (Fig. 3.15) in the floor of the wider valley and provided an excellent site for a dam. However the rocks of the area consisted of alternating clays and highly porous limestones, neither of which were particularly strong. A period of heavy rain saturated the porous rocks and the slide, 1.9 km long and up to 200 m thick, started to move slowly. On the night of 9 October 1963 the movement accelerated and in approximately one minute the mass slid into the reservoir, completely filling part of it and creating waves rising 250 m above the original lake level at one point. Two thousand people lost their lives when the wave swept down the 5 km long reservoir, overtopped the dam and inundated the village of Langarone.

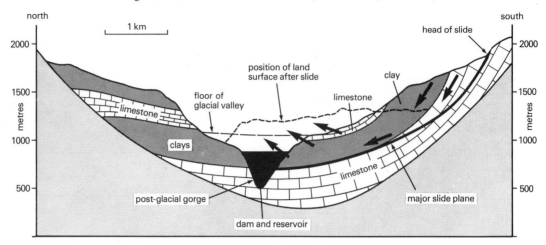

Fig. 3.15 The Vaiont Dam rock slide (After Kiersch)

The saturation of the limestones increased the weight of the rocks above the slip plane and provided lubrication by raising the pore pressure (page 49) at the slip surface. The rise of the water table caused by the build-up of the level in the reservoir tended to 'float' the lower part of the slide. The water in the reservoir had no retaining effect on the slide. The coefficient of friction was reduced to a value low enough to allow the rock mass to slide rapidly down the gently inclined slide plane.

Slides on a much smaller scale are quite common. Figure 3.16 shows a slide on a embankment at a motorway access point. The thin turf layer and soil absorbed water until it became too heavy and tore the turf near the top of the slope, the mass then slid easily over the wet surface of the consolidated boulder-clay fill beneath. At the foot of the slope the mass can be seen to have broken up and almost flowed onto the hard shoulder.

Fig. 3.16 Field sketch of a slope failure on a motorway embankment

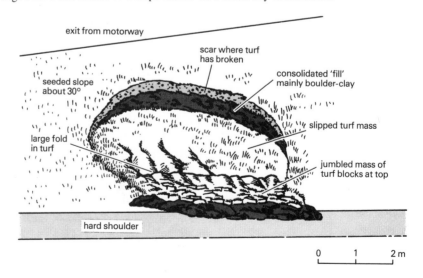

Fig. 3.17 Sketch profile of a slump and flow on the north Norfolk coast

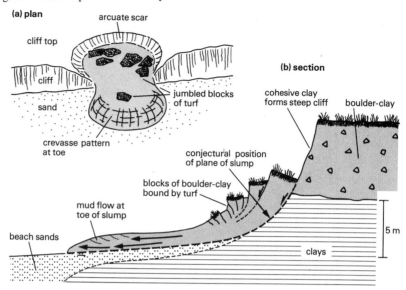

(a) plan

arcuate scar

cliff top

cliff

sand

jumbled blocks
of turf

crevasse pattern
at toe

(b) section

cohesive clay
forms steep cliff

boulder-clay

conjectural position
of plane of slump

blocks of boulder-clay
bound by turf

mud flow at
toe of slump

beach sands

5 m

clays

3 Slumps

Slumps usually occur in weaker rocks than slides and are distinguished from them by having a *rotational* movement along a curved slip plane (Fig. 3.14).

The slump shown in Fig. 3.17 was seen on the north Norfolk coast, where a chalky boulder-clay overlay a soft Pleistocene clay. Before the slump occurred heavy rain and gales had saturated the rock and undercut the cliff. The slump was triggered off by the clay flowing away from the foot of the cliff, and the boulder-clay above slumped onto the clay when its mass was unsupported. Large blocks of the boulder-clay were transported on top of the flowing clay.

Small scale slumps occur on debris-covered slopes which have a vegetation cover. These *terracettes* rotate about an imaginary point producing a change in the angle of tilt of the vegetation mat (Fig. 3.18). The breaks in the vegetation mat, the 'risers' mark the exposed positions of the possible slip planes. An

Fig. 3.18 Terracettes

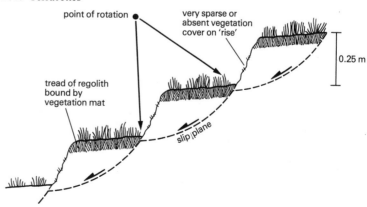

point of rotation

very sparse or
absent vegetation
cover on 'rise'

0.25 m

tread of regolith
bound by
vegetation mat

slip plane

alternative explanation of the terracettes which occur on slopes of 18° and steeper, is that they are due to the trampling of animals who traverse the slope almost parallel to the contours using these well defined 'paths'. This is supported by the presence of 'ramps' which lead from one level to the next and form an interconnecting network of paths. The explanation of these 'sheep' tracks' probably lies in a combination of the two.

Besides movements in which there is a *shear*, i.e. the surface of one unit moving relative to the surface of an adjacent unit, creep occurs by *deformation*, a change in shape without noticeable shearing. Clayey and silty regoliths change shape by minute movement between very small particles throughout the regolith. Consequently, shear movements occur along localised planes and deformational movements are dispersed.

With a constant applied force, clay deformation continues steadily. This is dependent on the water content of the clay. If it is dry, it behaves as a solid. The minimum water content at which this type of deformation is possible is called the *plastic limit*. The plastic limit is defined as the minimum water content which will allow the clay to be rolled into a thread 3.175 mm (1/8 in) in diameter. Clays may deform as plastic bodies.

With a greater water content the clay will deform under its own weight. This type of movement is called *flow* and the clay changes into a liquid (Fig. 3.14). The minimum water content at which this type of movement can occur is called the *liquid limit*.

Usually the regolith can be regarded as plastic. It is deformed by the slide component of gravity and remains deformed even when the force is removed. *Creep* by deformation is important in clayey regolith materials (page 58).

4 Flow

Slope materials with a high proportion of fine particles — more than 35 per cent by weight — are prone to movement by flow. Previously flow was defined as movement of a mass by internal deformation under its own weight. This property is dependent on the clay becoming saturated with water to a percentage greater than the liquid limit. The clays at the base of the cliff in Fig. 3.17 flowed away from the foot of the slope leaving the boulder-clay above unsupported and causing it to slump.

Flows are generally faster than creep (page 58), although less persistent. In many cases the toe of a slump changes to a flow as the water content rises and the nature of the material becomes finer. The slopes on which flows occur can be gentle, 5–6°, but most originate on slopes of 10° or more and flow to gentler areas. The flow in Fig. 3.17b stagnated on the beach with a slope of 3–4°. It spread out into a lobate 'foot' and as it dried out it developed a series of radial and concentric 'crevasses' reminiscent of a piedmont glacier (page 55).

(i) The Aberfan *earthflow* shown in Plate 3.4 originated from a saturated colliery spoil tip situated at the top of a 200–300 m valley side slope of about 14°. Heavy rain raised the water content of part of the tip above the liquid limit and the flow began to move. It rapidly gained momentum and overwhelmed the school at the lower concavity of the slope. Here the lower slope allowed water to drain from the flow and its velocity fell to near zero, while the

Plate 3.4 Earthflow at Aberfan, South Wales (Western Mail)

upper parts continued to move down into the village. The most difficult and urgent part of the rescue work was to construct ditches to drain the water from the flow, reduce its water content below the liquid limit and cause it to 'set'. In order to prevent a repetition of this disaster, the National Coal Board maintains routine drainage of tips and has removed many from precarious slope crest positions.

(ii) *Mudflows* are more rapid, less viscous and flow on much lower slopes than earthflows. They occur most commonly in areas with very sparse vegetation cover and subject to torrential downpours. The exposed regolith rapidly becomes saturated, exceeds its liquid limit, and in effect becomes a viscous river. Mudflows in desert wadis have been reported up to 2 m thick and moving so fast that they had waves on their surface. They have usually stopped on lower angle slopes where the water drains from the base of the flow into permeable regolith beneath.

(iii) The word *solifluction* literally means soil movement but is most frequently used to describe movements under tundra and subarctic conditions. In summer

the layer above the permafrost melts and becomes mobile. This is called the *active layer*. The water content is very high because the permafrost prevents downward drainage of meltwater. Flow may then occur in the active layer distorting the vegetation mat. These flows form lobes 10–50 m wide and up to 2 m high (Plate 3.5) in a series of step-like terraces down the slope.

The rate at which solifluction moves material on the slope varies from slope to slope and between different places on the same slope. Solifluction in Britain today is found in mountains about 1000 m high in the Cairngorms, and many fossil solifluction deposits are known, as in the chalk downs of southern England, which experienced periglacial conditions during the Pleistocene glaciation (page 202). The landforms associated with solifluction are dealt with in Chapter 8 on Periglacial environments.

5. Soil creep

Creep is a process which goes on more or less continuously in regolith materials. It contrasts with falls, slides and slumps which are events occurring rapidly over very short periods of time.

Most slopes steeper than about 6° show signs of creep. Figure 3.19 illustrates some of the more obvious effects of creep. The build-up of debris behind walls is due not only to creep but also to slope wash (page 62). Tilted pylons and telegraph poles result from their being sited on inadequate foundations which do not reach down as far as the solid bedrock or the stable part of the regolith. The downward movement of the regolith may cause tension gashes in roads with poor foundations, when the outside edge of the road tries to travel faster down the slope than the inner edge. The *curvature* of the exposed ends of well bedded or cleaved rocks shows that creep affects bedrock in a relatively

Fig. 3.19 The effects of soil creep

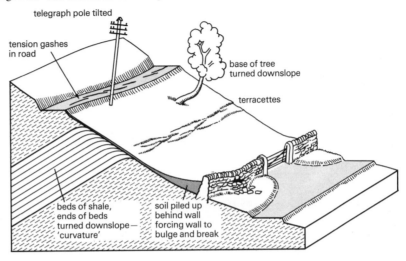

unweathered state as well as true regolith. The creep effectively 'plucks' the rock fragments from the bedrock. Terracettes, previously described under the process of slumping, may equally well be ascribed to creep, where they behave like the treads on an escalator. They may be initiated by a rotational slump effect but are then carried downslope maintaining a stable angle to the slope.

The mechanism of soil creep comprises a number of physical processes. The first of these is *expansion and contraction* of the regolith. This may be caused by *wetting and drying*, which causes clays, particularly montmorillonite, to expand considerably, or by *heating and cooling*, which causes all particles to change their volume. Water in the regolith expands as it falls below freezing point and in humid cold areas this is possibly one of the most effective forces in causing expansion of the regolith.

Expansion causes the surface of the slope to *heave* at right angles to the original surface and particles are lifted along the *expansion path* (Fig. 3.20).

Fig. 3.20 Movement on slopes caused by expansion and contraction

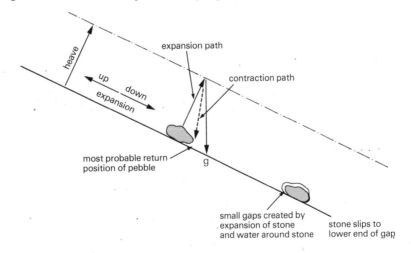

Gravity and the contraction of the regolith around the particle affect the return path, the most probable position for the particle on its return is between the lines representing the directions in which these forces are acting. The net result is a downward movement of this and every particle on the slope.

The second process involves particles expanding linearly, parallel to the slope. On expansion, the smaller fragments are pushed aside, leaving a space around the larger fragment. If the slope is steeper than the angle of friction for the material, the fragment slides to the lower end of the space (Fig. 3.20).

Third, the expansion parallel to the slope pushes down the slope and adds to the *slide* component of gravity, thus promoting the movement of material which lies near to its angle of friction.

Freezing of saturated regolith sometimes produces a growth of *ice needles* perpendicular to the surface called *pipkrakes*. These have been observed to lift loose surface layers of debris up to 20 mm thick over 100 mm above their original position. When the pipkrakes melt, the debris is let down vertically or often falls or rolls down the slope. In Britain, unvegetated slopes with a high water content and a significant silt or clay content, e.g. boulder-clays, are frequently affected by pipkrakes during winter frosts.

Fig. 3.21 The velocity profile of a creep slope

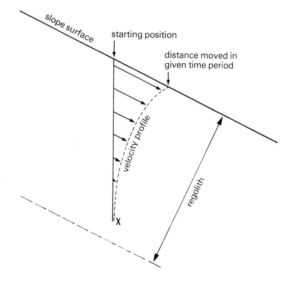

The process of creep affects the top metre of the regolith most severely; Fig. 3.21 shows how the rate of movement decreases rapidly with depth until it reaches zero at X. Usually there is no basal slide component of movement at the base of the velocity profile, otherwise this would constitute a slide and movement would be more rapid.

(a) Measuring movements on creep slopes

The methods of measuring the velocity profile are illustrated in Fig. 3.22. The surface velocity is most easily measured by inserting pegs and measuring the distance between them and fixed reference points on bedrock or buildings after an interval of time.

'T' bars are simply metal T-shaped rods, 0.3–0.4 m long and about 0.1 m across the top of the T. They are inserted into the ground and the angle of tilt of the top recorded. As the regolith moves, the bars become progressively more tilted, enabling the rate of movement of slopes to be compared.

A length of dowel, sawn into sections 20–50 mm long and inserted as a column into a vertical hole in the soil, registers movements by deviating from the vertical. To observe this, a pit must be excavated with one face parallel to the direction of maximum slope.

Young pits entail the excavation of a pit to install horizontal metal bars in a vertical line down the wall. The bottom of the line is referenced to a bedrock marker and the pit is filled in. On re-excavation, the movements of the rods relative to the fixed point can be measured and the velocity profile constructed.

Fig. 3.22 Methods of measuring slope movements

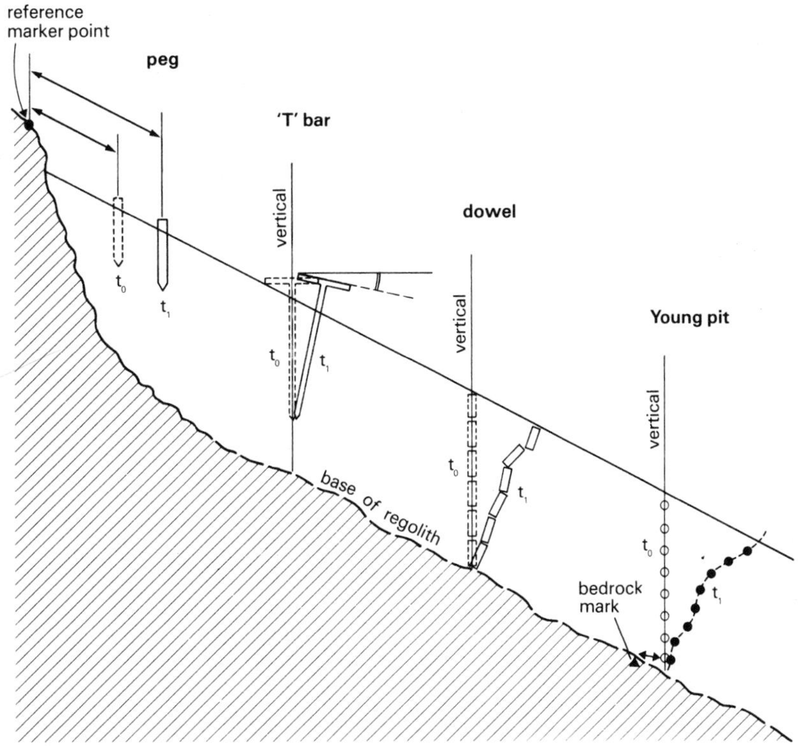

With all these methods some disturbance of the slope is inevitable. It is most severe in cases where a pit is dug. Many people would argue that this invalidates the measurements, but there are few practical alternatives.

Measured rates of soil creep are usually quite slow. In temperate humid climates on a 10° slope the creep rate is only 1–2 mm/annum. Much higher rates are found on slopes where there is no vegetation cover or high water content. These movements, however, are occurring continuously and over the entire slope. In total their effect is far greater than the sporadic movements of slumps and slides.

Because of this slow movement, creep studies are long term. The movement in one year is often less than the measuring equipment is able to resolve.

6. Surface wash

Unlike the other types of slope movements, slope wash is not a mass movement phenomenon. It occurs on slopes when either the intensity of rainfall exceeds the capacity of the soil to absorb water, the *infiltration capacity* (page 79), or where the water table lies at the surface. Figure 3.23 illustrates how both prevent infiltration of water into the soil, and hence promote flow of the excess *over* the surface. This is termed *overland flow*.

Fig. 3.23 Wash slopes

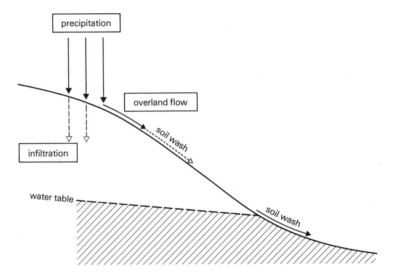

(a) Sheet flow

Frequently, overland flow occurs as a thin continuous layer called *sheet flow*, which washes particles, mainly silt- and clay-sized, down the slope. This erosion occurs when the flow of water has reached a critical velocity and is powerful enough to overcome the inertia of the particle and the *resistance* or *cohesion* of the soil (page 106). The lower part of the slope receives water draining from the slope above. Hence the volume of water passing over the slope is proportional to the length of the slope above the point at which the flow is measured. The greater volume of water lower down the slope transports larger volumes of debris more efficiently and this may be responsible for the lower concavity of slope profiles.

The effect of sheet flow is diminished drastically by the presence of vegetation to the slope. It reduces the velocity of flow by forming miniature dams and diversions. The soil surface is bound together mechanically by roots and by the formation of humus which, together with the clay minerals, forms cohesive units of soil particles.

The vegetation cover is dependent on factors of climate; sheet wash is far more effective in arid and semi-arid areas, where rainfall intensity is generally high even though total amounts are quite low and rain is an infrequent event. In temperate humid areas, estimates of the rate of lowering of the surface by sheet wash vary from 0.01 mm/a down to zero where vegetation is dense. In tropical rain forest, the rate of 15 mm/a has been quoted. Evidently there can be a very large variation even on different parts of the same slope.

(b) Raindrop impact

Raindrops in an intense storm have an immense amount of kinetic energy. Individual drops rarely exceed 2 mm in diameter and these have a terminal velocity of 6.9 m/s; in a storm, with a rainfall intensity of 100 mm/h, one spot might be hit on average 23 times per hour. When these drops hit the soil surface they cause a splash of soil particles which may be up to 0.6 m high but is usually only 0.3 m above the ground surface. On the flat these particles merely change position (Fig. 3.24a), but on a slope there is a net downslope

Fig. 3.24 Splash erosion

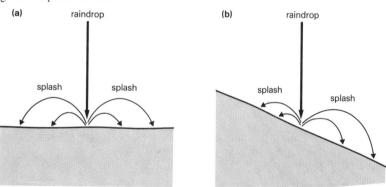

transfer since those particles splashed towards the lower part of the slope have slightly longer trajectories than those which move upslope from the point of impact (Fig. 3.24b).

Raindrop impact has been measured to move 225 tonnes of soil per hectare in 1.25 hours of heavy rainfall. The effect of this is to leave particles of sand size and larger on the slope, while removing the finer particles to the base of the slope.

In sheltered areas, where the rain does not often fall at the angle as it is blown by the wind, large pebbles protect the soil beneath from rain splash and an earth *pillar* forms when the unprotected soil around the pebbles is splashed away. Plate 3.6 shows earth pillars in a cave where drip from the roof has had the same effect as rain. Bare soil surfaces under trees often have earth pillars.

Both splash and wash are particularly effective on unvegetated slopes under cultivation and are major causes of soil erosion where intense downpours are common, such as in the interior lowlands of the USA.

Plate 3.6 Earth pillars in
Dan-yr Ogof, South Wales
(A.S. Freem)

Measurement of surface wash

It is possible to measure surface wash by assessing either the *erosion* or the
deposited sediment. Rods or nails inserted into the ground can be measured
periodically to determine if the surface around them has risen or fallen. The
distance from the top of the nail is more easily measured if a washer is placed
on the nail when it is inserted. This falls as the surface is lowered but is buried
by deposition. This method measures all types of surface erosion.

Sediment traps of various designs have been used to catch material moved by
wash. These consist of a covered box with a lip flush with the soil surface.
When the process of wash is in operation sediment enters the box from upslope
and settles in the still water in the box; the excess water either spills around the
edges or out of the special outlet (Fig. 3.25). The weight of the sediment in a

Fig. 3.25 A slope sediment trap

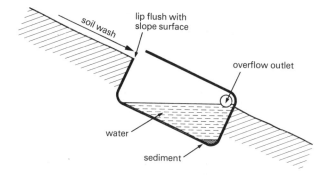

dry state divided by the length of the lip in metres gives the sediment yield in grams per metre of the slope width.

ASSIGNMENTS

1. (a) *Look at the graphs in Fig. 3.26. Which type of movement from slide, creep and flow does each represent?*
   (b) *What time interval might you expect between t_0 and t_1 in each case? Describe the factors which determine these time intervals.*

Fig. 3.26 Velocity profiles of slope movements

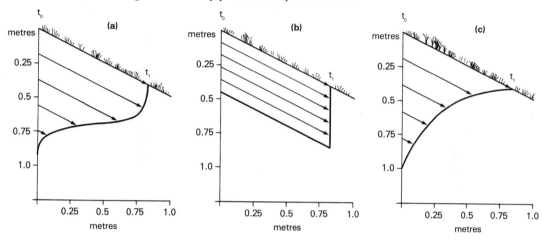

2. *A boulder-clay covered slope is gently concave with a curvature of $-6°/100$ m. In plan the slope is water gathering. Describe the sequence of events which might lead to an earthflow occurring on the slope. In the description you should include the terms liquid limit, infiltration, flow, plastic limit, angle of internal friction, pore pressure, lubrication, deformation, scar, toe, lobate and precipitation.*
3. (a) *From the data in Table 3.3, draw a velocity profile of the slope regolith movement.*
   (b) *If a T bar had been inserted 0.4 m vertically into the regolith, to what angle would it have been tilted in the period that the velocity was measured, assuming that it pivoted about its lower point? From this, what errors can you see in the measurement of slope movement using the T bar method?*

Table 3.3 Movement measured on a slope

Depth (mm)	Distance moved (mm)
10	16
100	11
200	4
300	1
400	0

Plate 3.7 Folkestone Warren landslips (Cambridge University Collection — copyright reserved)

Table 3.4 A classification of mass movements

		Mass movements		
		← ——— Increasing water content ← ———→		
		Flows	*Plastic deformation*	*Shears (fracture)*
Rocks	Rapid ↓ Slow	· · · · · · · · ·	· · · · · · · · · · · · · · · ·	· ·
Regolith	Rapid ↓ Slow	· ·	· · · · · · · · · · · · · · · ·	· ·

4. *Plate 3.7 shows part of a large series of slope movements on the Channel coast near Folkestone. Make an annotated sketch to show the main features resulting from and responsible for these movements.*
5. *Describe the processes operating on a free face which cause failure of the rock and rockfall (page 49).*
6. *Complete the classification of mass movements in Table 3.4. The missing terms are creep, earthflow, slump, debris (earth) slide, solifluction, rockfall, mudflow, rockslide and rock avalanche. (Some of these terms apply both to rock and regolith materials.)*

D. Slope Development

The slope processes examined in the previous section may be thought of as a conveyor belt, transporting the products of weathering on the initial part of their journey to the sea. The lower end of the conveyor usually dumps its load into a stream channel and it is rapidly taken away. If this does not happen, the conveyor grinds to a halt, the debris accumulates and weathering slows down as a protective mantle builds up. This simple analogy illustrates the basic role of the *slope system*, that of *transport*. There are, of course, a large number of intermediate states between efficiency and stagnation. An examination of some of these states in theory illustrates some of the ways a slope may develop.

1. States of the slope system

In Fig. 3.27a, a rock slope is covered with a thin regolith and the *input* into the segment is exactly balanced by the Fig. 3.27 *output*. If these conditions remain constant over a period of time, the angle of slope and the volume of regolith in *store* on the slope will not change. However, weathering within the segment is contributing material to this store and acting as a secondary input into the ever

Fig. 3.27 States of the slope system

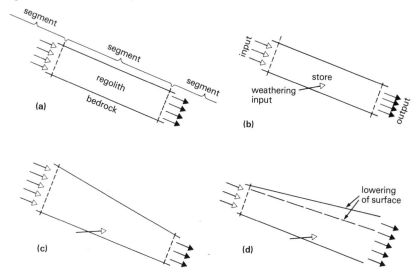

67

moving stream of slope debris. In order to maintain the slope angle and keep the store volume constant the segment must have an increased output, as in Fig. 3.27b.

The volume of regolith passing through a slope segment in a given time interval increases towards the base of the slope. Each successive downslope segment must accommodate the input of the previous segment plus the products of weathering from that segment in order to keep angle and regolith thickness constant. In order to achieve this the transport rate on the slope must increase (Fig. 3.27b). W.M. Davis called this steady state with a constant regolith thickness a *graded slope*. No bedrock shows through and the slope appears to be in equilibrium with the climate, weathering and the rates of slope movement.

The upper convexity (page 36) was explained by G.K. Gilbert in this way. He proposed that the transport rate was increased by a steepening of the slope angle, and that the volume of debris transported was correspondingly increased downslope.

If the input exceeds the output as in Fig. 3.27c, the volume of material in store increases and the regolith at the top of the segment thickens. This indicates that the transport system operating on the slope is working at a maximum. Further development of the slope is limited by the transport rate. The slope of the surface then becomes steeper than the bedrock slope. This causes an increase in the velocity of slope movement which tends to lower the slope angle by transporting more material to the bottom of the slope and effectively restores the balance between input and output. A scree slope fed by rapid stone fall from a cliff will steepen until there is a tendency for the scree to exceed the angle or repose of the scree. The transport rate rapidly accelerates and boulders slide, fall and roll, lowering the slope angle and restoring the stability of the scree.

Where the output is greater than the input the volume in store falls (Fig. 3.27d), the regolith thins as debris is moved away downslope and, if the process continues, eventually bedrock will be exposed. The development of this type of slope is impeded by the rate of weathering since the transport system can carry all the material supplied to it.

The base of the slope usually 'outputs' debris into a stream channel (page 126) or into the sea (page 241). Both these agents are very effective in removing the debris faster than the slope can supply it. This is termed *unimpeded basal removal* and it tends to accelerate slope movements by removing resistance to downslope movement from the last segment. The effect is fed back up the slope from segment to segment eventually affecting the whole profile.

Should the removal of debris from the base of the slope be inadequate to keep up with the rate of supply from the slope, then case (c) in Fig. 3.27 operates and the feedback slows down the rate of movement as the regolith accumulates at the bottom of the slope. These are situations of *impeded basal removal*.

Real slopes are rarely as regular as those shown in Fig. 3.27. The relationships between input and output vary on different parts of the slope. The four cases described help in the understanding of the ways slopes develop over time. The theories of slope development extend these principles to whole slopes and thus to the total landscape.

2. Theories of slope evolution

Speculation about the ways in which slopes evolve have been regarded as very important by geomorphologists since the latter part of the nineteenth century. The evolution of slopes provides clues as to how the landscape as a whole has evolved. The debate has continued for so long mainly because the period of time required to produce a significant change on slopes is itself immense (page 52). All theories are of necessity dependent on evidence preserved in the landscape. The measurements made on slopes have taken place over such a short period in relation to the age of the slope that they have thrown little light on the true mode of their evolution.

There are three widely accepted models of slope evolution: slope decline, slope retreat and slope replacement.

(i) Slope decline

Slopes which are evolving by *slope decline* show a progressive decrease in the angle of slope in each phase of their development (Fig. 3.28a). The elimination

Fig. 3.28 Models of slope evolution

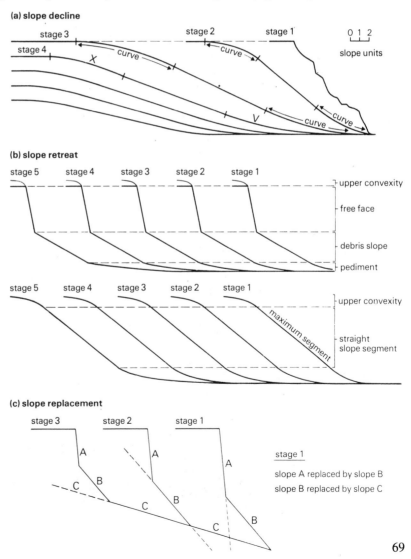

(a) slope decline

stage 3

stage 4

curve

curve

stage 2   stage 1

curve

curve

0 1 2

slope units

(b) slope retreat

stage 5   stage 4   stage 3   stage 2   stage 1

upper convexity

free face

debris slope

pediment

stage 5   stage 4   stage 3   stage 2   stage 1

maximum segment

upper convexity

straight slope segment

(c) slope replacement

stage 3   stage 2   stage 1

A

B

C

A

B

C

A

B

C

stage 1

slope A replaced by slope B

slope B replaced by slope C

69

of the free faces in stage 1 is by the processes of fall and slump of the bedrock until the slope is gentle enough to develop a cover of regolith. The slope will then be less steep than 35° (page 38). Stage 2 shows this phase which is called the *graded slope*. Case (b) in Fig. 3.27 is dominant over the majority of the slope, therefore the regolith maintains a constant thickness over the slope and all the weathered material is transported by mass movements and wash. The form of the slope is concavo-convex (page 36). The curvature of these elements decreases as the slope continues to decline, stage 3 and 4 in Fig. 3.28a, and the length of the straight segment diminishes. The upper convexity corresponds to case (d) and the lower concavity to case (c) in Fig. 3.27. The possible origin of each of these elements was discussed on pages 62 and 68. W.M. Davis, the original proponent of this theory, based his arguments on visual assessments of slopes in humid temperate areas, particularly the Appalachians in the north-east of the USA; concavo-convex profiles are common in such areas.

(ii) Slope retreat

The form of *retreating* slopes is shown in Fig. 3.28b. The free face (page 36) may be absent from a particular slope but all the elements except the *pediment* maintain a constant length and angle as the slope develops. The pediment is the name given to the gently concave area which extends from the foot of the debris slope and becomes wider and wider as the slope retreats. In Plate 3.8 the pediment extends off the photograph for several miles in each direction to the

Plate 3.8 Hill slopes in Morocco (Author's photograph)

foot of another small hill like that shown. The pediment is generally slightly concave but the slopes are very gentle, only 3–5°, and often they appear to be completely flat.

The angle of the free face at the top of the slope is determined by the strength of the rock. A strong rock, resistant to chemical weathering, will often form a vertical free face. In less resistant rocks, the free face may not be present (Fig. 3.28c). Rapid weathering reduces the slope to a continuous debris-covered slope. In many semi-arid areas a cap rock of laterite forms the free face. The origin of this rock was explained on page 26, and Plate 3.9 shows this situation in Australia. All these slopes are thought to retain a constant angle throughout their development. The debris on the slope maintains a

Plate 3.9 A small residual hill in Western Australia (I. Kaill)

constant thickness indicating that inputs and outputs on this part of the slope are in balance. This corresponds to case (b) in Fig. 3.27.

The pediment is covered by only a thin sheet of regolith. We must conclude that slope debris reaching the pediment is all carried away, case (d) in Fig. 3.27. This lack of debris in semi-arid areas is difficult to explain. Rivers are spasmodic and there is little evidence to indicate the process which could be responsible for removing such large volumes of material.

(iii) Slope replacement

The slope replacement theory envisages original steep slopes being replaced by lower angle slopes which extend upwards from the base at a constant angle. In Fig. 3.28d a free face slope, is slowly buried by a scree which accumulates at the base of the cliff. In stage 2 the scree is replaced by B, a lower angle slope perhaps formed by small particles being washed from the scree. This slope is in turn replaced by an even lower angled slope, C, which again extends upwards from the base. In contrast to retreating slopes there is a change in the length of the segments as the slope develops; some, like the free face may even be eliminated.

Walther Penck explained these ideas in the 1920s. Debate about his ideas and those of others presented here still continues. Experiments and observation of slope processes have done little to resolve the problem and evidence in the landscape is still of great importance.

3. Evidence of slope evolution

Dury has illustrated from Northamptonshire how slopes developed on Lias clays have a characteristic angle of 11°, which probably represents the

Fig. 3.29 Evidence of slope evolution ((a) After Dury and   (b) After Savigear)

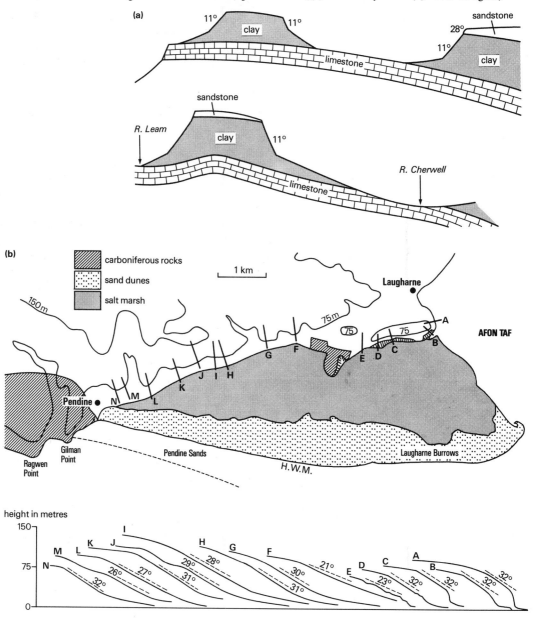

'strength' of the clay. This angle is the maximum angle at which the slope is stable when considered over a long time period (Fig. 3.29a). Even where the slope is capped by a sandstone the clay maintains its 11° angle. These slopes at their base merge into lower angle 'pediment' slopes which in the same area pass smoothly across a geological boundary onto Lias limestones.

The 11° slopes appear to have reached their present position by parallel retreat, and the pediment is smoothly concave. These slopes do not have strong basal removal, they are controlled by the operation of slope processes and are transport impeded (page 68).

In South Wales, at Pendine, a cliff of Devonian sandstone has been progressively isolated from the sea by a spit which has grown eastwards from Gilman Point (Fig. 3.29b). The western slopes were isolated from the sea when the spit first began to form and eastern slopes are still subject to wave attack and rapid basal removal of slope debris. The slope profiles therefore represent a sequence of slope development with longer periods of impeded basal removal towards the west.

The 'youngest' slope, A, has a bevel at 32° above the free face and this characteristic angle persists in the slopes through to N and occurs in slopes further west than this (Fig. 3.29c). There is little significant difference between the maximum segments and so it seems safe to assume that they retreat according to the parallel retreat theory.

At the foot of the slope, regolith material is accumulating in the absence of wave action. The slope is usually concave and is replacing the slope above, conforming to the replacement theory. There is some evidence here to suggest that the slopes with impeded removal slowly decline with age. The maximum segments decline gently towards the west.

This evidence seems to indicate that the three theories can operate together in one area. Each process may be operating on the slope in different places; no right or wrong theory exists. There is still no unequivocal evidence to solve the 'slopes debate'.

ASSIGNMENTS

1. *Table 3.5 represents slope profiles of stage 3 in the hypothetical slopes shown in Fig. 3.28.*
   (a) *Construct a slope histogram for each slope. Compare the form of each of these histograms.*
   (b) *Explain how it might be possible to distinguish between the modes of slope evolution using these methods (see page 38). Which of the histograms most closely resembles that constructed from Table 3.2? How does this assist in explaining the form of that slope?*
   (c) *What changes would occur in the histograms if they were constructed for later stages in the development of the slopes?*
2. *Explain how 'basal removal' (page 68) can influence the processes operating on a slope.*
3. *Compare the states of the slope system which would be found on a quartzite, a resistant unjointed rock, and on a shale, a relatively soft well-bedded rock.*

Table 3.5 Slope histogram data

Profile 1	Profile 2	Profile 3
00	00	00
03	03	48
07	46	49
10	46	49
14	46	50
19	46	36
21	42	37
24	40	37
24	38	36
25	37	36
24	37	17
24	36	17
25	36	17
24	16	16
22	06	17
16	05	17
13	02	09
10	02	08
08	01	08
07	01	08

Key Ideas

A. *Introduction*
1. Contours, which are rarely absent from topographical maps illustrate the widespread and important nature of slopes.

B. *Slope form*
1. The *slope angle* is measured with reference to the horizontal and it is determined by the horizontal length of the slope and its vertical height.
2. *Slope segments* are portions of the slope where the angle is constant and therefore the slope is straight. *Slope elements* are curved portions of the slope profile. The *curvature* is expressed in °/100 m of slope.
3. The majority of slopes are less than 45°.
4. Frequently occurring slope angles are called *characteristic angles*.
5. Slopes in plan may be *straight, concave or convex*; each of these is characteristic of particular locations within a river basin.

C. Slope processes and landforms
1. Particles are retained on slopes against the force of gravity by frictional forces. The angle at which a particle begins to move is the *angle of friction*. The tangent of this angle is the *coefficient of friction*, the ratio of the *slide* and *stick* components.
2. Water lubricates the contact between particles in slope regoliths and *pore pressure* reduces the effective forces between them.
3. Rocks which are strong or cohesive require forces greater than that of gravity to cause them to shear.

4. Mass movements with small or negligible *internal deformations* and straight lines of motion are *slides*.
5. *Rotational motions* with negligible internal deformation are *slumps*.
6. *Creep* is a movement of regolith by internal deformation and limited shear. It is the product of expansion, contraction, freezing and thawing combined with the effect of gravity.
7. *Plastic deformation* occurs when an applied force brings about a constant and steady change in shape. The *plastic limit* is the minimum water content at which plastic deformation will occur.
8. The *fines*, silt- and clay-sized particles, determine the water retaining capacity of the regolith.
9. *Earthflow* and *mudflow* represent progressively faster and wetter types of flow movements.
10. *Sheet flow* and *rain splash* move individual particles downslope and therefore are not strictly mass movements.

D. *Slope development*
1. The relationship between input and output on a slope determines whether the slope will become steeper or more gentle.
2. Perfect balance between input and output results in a *graded slope*.
3. Slopes change in time by *declining, retreating* or *replacing* the original slope. These changes may operate in conjunction on one slope.

Additional Activities

1. To test the hypothesis that rock types have characteristic angles of slope under the same climatic conditions:
   (a) Measure a series of profiles using the methods on page 39 on different rock types in your area.
   (b) Construct slope histograms and/or slope frequency graphs (page 46) in order to compare the slopes. Students' 't' test can be used to examine the significance of the difference between the means of the slope angles on the various rock types.
   (c) Profiles on the same rock types may show variations with aspect. It is possible to test this hypothesis using the same techniques.
2. In section C it was suggested that: (i) screes were sorted with the smallest particles lying highest on the slope (page 52), and (ii) that the curvature of scree profiles varies with the height of the free face which feeds the scree. It is possible to test (i) by selecting sample points on screes and measuring a sample of 20–50 stones at each, using the methods on page 116. Analysis of the data collected could be a plot of the mean size of boulders at each point against distance from the top of the slope.

   If several scree profiles are surveyed and the curvature of each plotted against cliff height, it is possible to test the validity of (ii).
3. Explain how the water content of a regolith or rock slope may influence mass movements.
4. Describe the ways in which weathering and lithology influence slope processes.

Plate 3.10 Hill slopes in the Howgill fells, North Pennines (Author's photograph)

5. Plates 3.9 and 3.10 show hill slopes developed under different climatic and lithological conditions. Compare the differences in form shown by these hills. Discuss the processes responsible for these forms under the different climatic conditions.

Reading

YOUNG, A. and YOUNG, D.M., *Slope development*, Macmillan, 1974

WALTHAM, A.C., *Catastrophe: violent earth*, Macmillan, pages 49–79, 1978

LEOPOLD, L.B., *et al.*, *Fluvial processes in geomorphology*, W.H. Freeman, Chapter 12, 1964

*YOUNG, A., *Slopes*, Longman, Chapters 4–8, 13, 1975

*COOKE, R.U., and DOORNKAMP, J.C., *Geomorphology in environmental management*, Oxford University Press, Chapter 6, 1974

*BRUNSDEN, D., *Slopes: form and process*, Institute of British Geographers, 1971

4 *Fluvial environments*

A. Introduction: the Hydrological Cycle

Running water is the dominant agent in producing changes in the physical landscape and its effects are widespread. The physical and chemical disintegration of rock and mass wasting (Chapter 2) supply material which is removed by rivers to the sea. Other agents, such as ice sheets and glaciers, the wind and waves, all make changes in the landscape. The effects of these will be considered in later chapters, where it will be seen how each may be important on a local or regional scale. In terms of overall effects, however, the work of rivers is often more significant than that of other agents, as the time scale over which they operate may be much longer.

A river or stream is part of an orderly sequence or cycle of events taking place within the hydrosphere — the *hydrological cycle*. Water from the oceans is evaporated and is transported by winds to the land where, under certain conditions, it falls as rain. In a variety of ways, including *run-off* via streams and rivers, the water finds its way back to the sea (Fig. 4.1).

Fig. 4.1 The hydrological cycle

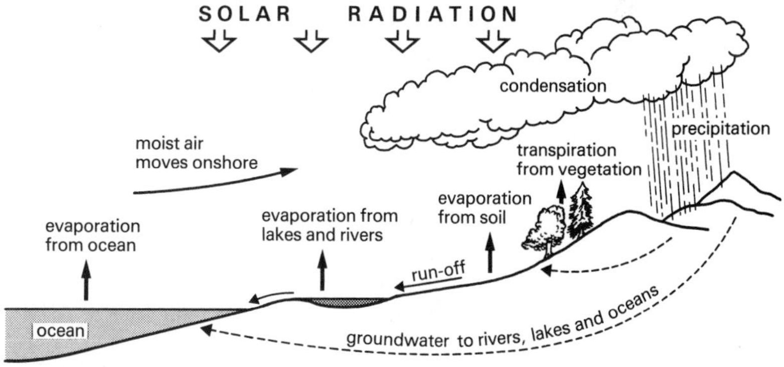

There are no effective gains or losses in this cycle since the atmosphere and lithosphere together have a fixed amount of water. Therefore, we can think of this as a closed system, as defined in the introductory chapter (page 11), energised by solar radiation.

B. The Drainage Basin

The effective part of the hydrological cycle in producing changes in the land surface extends from the point where the precipitation first hits the ground to where it re-enters the sea (Fig. 4.1). In its course over the land surface, water always takes the steepest downslope route available. This is indicated on a contour map (Fig. 4.2) by lines which cross the contours at right angles giving the greatest change in height in the shortest horizontal distance. Rain which

Fig. 4.2 A typical drainage basin

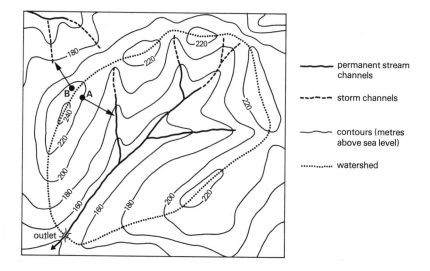

falls at A in Fig 4.2 flows down the slope to join the stream below, while that falling at B flows downslope to another stream on the other side of the hill. At the crest of the hill, between A and B, is an imaginary dividing line, the *watershed*, which separates the area in which water drains northwards from that in which water drains southwards. This line can be extended along the crests of the hills to separate the area draining to the river, which flows through the *outlet*, from areas which contribute their water to other rivers. The area enclosed by the watershed, and therefore draining through the outlet, is the *drainage basin* or *catchment area*. The whole of the land surface can be considered as a mosaic of drainage basins of varying size and shape. The drainage basin is the unit of the landscape with which this chapter is concerned.

Where slopes meet at the bottom of a valley, water is concentrated and forms a *river channel*. Some of these channels are *permanent*, as a result of a fairly constant supply of water from the slopes above, and some, *storm channels*, are *temporary*, only containing water during or immediately after rainfall (Fig. 4.2). Streams, therefore, do not begin at the watershed, since they need to be fed from slopes above. Water gathering slopes (page 44) concentrate water very efficiently and permanent streams often have their source at the base of such slopes.

1. The basin run-off system

A river basin is part of the hydrological cycle. The basin receives *inputs* and responds to these with a series of *outputs* and can therefore be regarded as an open system as defined in the introductory chapter.

The input refers to the incoming precipitation, mainly rainfall and snowfall, which falls within the watershed. This *infiltrates* the soil and is distributed through the system by a number of processes to become output. The main outputs are *evaporation, transpiration* and *run-off* (Fig. 4.3).

Fig. 4.3 A simplified diagram of the basin run-off system

precipitation

transpiration and evaporation from trees and plants

surface run-off

soil infiltration

evaporation from soil and rock surfaces

overland flow

storage of groundwater and flow

output

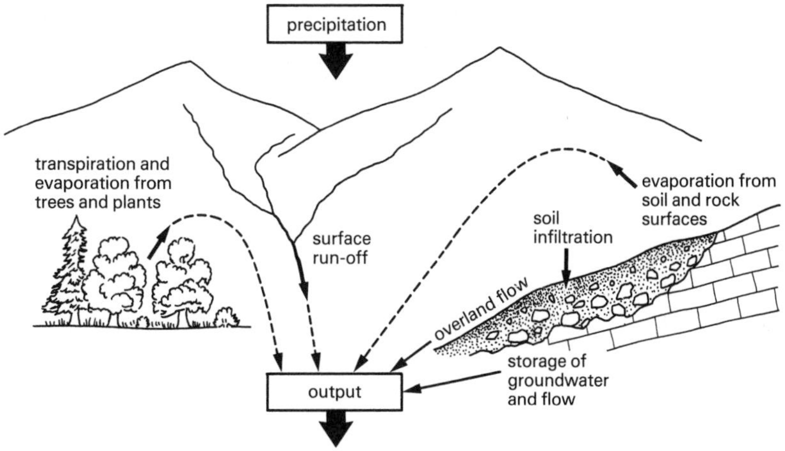

(a) Infiltration

Rainfall which reaches the ground surface does not immediately form streams and rivers. Part of it infiltrates the soil and rocks beneath the surface to be released into the river later. The amount of infiltration depends on how *permeable* the soil and rocks may be. The rate of infiltration of water into the soil changes with time (Fig. 4.4). At the onset of rain this rate can be quite high

Fig. 4.4 Infiltration rates and rainfall

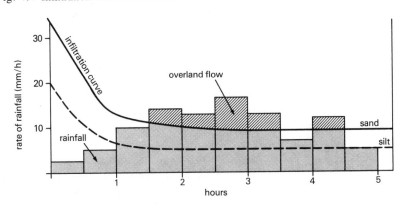

because the soil is dry. After a relatively short period of time the air spaces or *voids* between the soil particles become filled with water and the infiltration rate will be reduced to a lower but constant value. If the voids or pore spaces are too small to allow water to pass, the soil is *impermeable*. The small voids between clay and silt particles reduce the infiltration rate. In comparison, sands and gravels have high infiltration rates as a result of the large voids between the particles. The water passing through the soil is termed *throughflow*.

Water introduced into the underlying rock, *groundwater*, may be stored there for a relatively long period. This becomes a source of much stream flow during prolonged dry periods.

(b) Evaporation

Water is evaporated from the soil and from rock surfaces, and from the surfaces of leaves and branches which *intercept* the rain before it reaches the soil surface. The surface of the streams and lakes in the basin also provide an evaporation *output*.

(c) Transpiration

Vegetation removes water from the soil and evaporates it through the leaves. This output is termed *transpiration*. Evaporation and transpiration are difficult to measure separately and they are frequently referred to together as *evapotranspiration*.

(d) Run-off

Run-off refers to all the water which enters streams and leaves the basin as stream flow. Throughflow, through the soil, and groundwater flow both contribute to run-off. During a rainstorm the rate of rainfall may exceed the rate at which the soil is able to absorb the water, i.e. the infiltration rate. This excess becomes *overland flow* (Fig. 4.4). This flows across the surface until it reaches a stream to become run-off.

(e) Water balance

From the above it follows that a '*water balance*' exists to link inputs and outputs. This can be thought of as an equation:

Precipitation = evaporation + transpiration + run-off.

In the long term this equation must hold, but for purposes of shorter term analysis and measurement we must take into account the fact that the volume of water stored as groundwater, soil water and in the channels may change, and hence the equation should read:

Precipitation = evaporation + transpiration + run-off ± changes in storage.

Because of the difficulties in measuring evapotranspiration and water storage, it is usual to conduct water balance experiments starting and finishing at the time of minimum storage. In Britain this is at the end of the summer, when evapotranspiration has been high and rainfall low. Accordingly, the *water year* runs from 1 October to 30 September of the following year. In other areas of the world, the time of minimum storage is at a different time of the year (Fig. 4.19).

(f) Precipitation measurement

Although precipitation measurements, using rain gauges, are in themselves relatively simple, one of the problems to be overcome is that the rainfall varies over the basin. It is possible to assess the pattern of variation over the whole basin from data collected at a few randomly located gauges. Figure 4.5 shows two of the main methods of calculating total rainfall for the basin using *isohyets* and *Thiessen polygons*.

Fig. 4.5 Methods of averaging rainfall

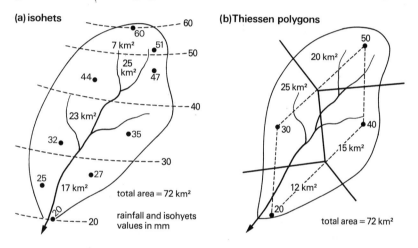

In the first method, isohyets are drawn from point values of rainfall. The areas between isohyets are calculated and the rainfall for each area calculated by taking the mid-isohyet figure multiplied by the surface area of each.

Thus the averaged rainfall in the example is:

$$\left(\frac{7}{72} \times 55\right) + \left(\frac{25}{72} \times 45\right) + \left(\frac{23}{72} \times 35\right) + \left(\frac{17}{72} \times 25\right)$$
$$= 38 \text{ mm}$$

In the polygon method, lines joining the positions of the rain gauges are bisected, dividing the basin up into polygons. The areas within each polygon are calculated. Then, as with the isohyet calculation above, the averaged rainfall in the whole basin will be:

$$\left(\frac{20}{72} \times 50\right) + \left(\frac{15}{72} \times 40\right) + \left(\frac{25}{72} \times 30\right) + \left(\frac{12}{72} \times 20\right)$$
$$= 47.2 \text{ mm}$$

1. (a) *Complete the flow diagram of the hydrological cycle in Fig. 4.6. The missing terms are: throughflow, evaporation, infiltration, transpiration, interception, soil storage, groundwater flow, overland flow, groundwater.*

Fig. 4.6 Flow diagram of the basin run-off system

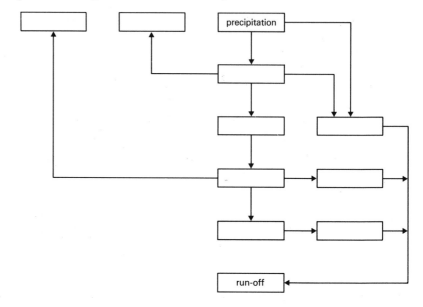

(b) *Why is it an example of an open system? (See page 11.)*

(c) *Write a short description and explanation of the hydrological cycle. Compare your description as a means of presentation with the above flow diagram.*

2. *Using a 1:25 000 map, select and draw on tracing paper a river and its tributaries, starting either from its mouth or estuary at the coast or from its confluence (joining point) with a larger river into which it flows. Use the method on page 78 to plot the boundaries of its drainage basin. Repeat this for neighbouring river systems, and hence build up a plot of several drainage basins. Do the boundaries of adjacent basins always coincide?*

3. (a) *Figure 4.7 shows a drainage basin with recorded amounts of rainfall at six rain gauge stations. Use both the methods explained in this section — isohyets and Thiessen polygons — to calculate the areal average of rainfall over the whole drainage basin. Your results should be the same or nearly so if you have been careful to be as accurate as possible.*

(b) *If evapotranspiration is equivalent to 20 per cent of the rainfall, and 10 per cent of the remaining precipitation infiltrates the groundwater table, what amount of rainfall reaches the stream channels? Express your answer in mm of rainfall.*

Fig. 4.7 Monthly rainfall totals for six stations

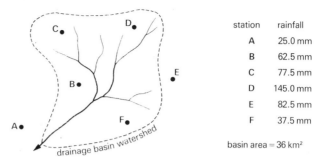

station	rainfall
A	25.0 mm
B	62.5 mm
C	77.5 mm
D	145.0 mm
E	82.5 mm
F	37.5 mm

basin area = 36 km²

(c) *If 4 per cent of the water reaching the stream is lost by direct evaporation from the channels what volume of water leaves the basin as runoff? (This calculation requires the area of the drainage basin.)*

2. Basin networks

River basins form complicated shapes on the earth's surface. In order to compare basins and to examine the effects of variables such as rainfall and vegetation cover, it is necessary to measure their shape. There are three basic properties which we will examine: linear properties — one dimension; areal properties — two dimensions; and relief properties — three dimensions.

(a) Linear properties

The linear parts of a river basin are the stream channels themselves. Water falling within the boundaries of a river basin eventually enters the stream channel, and the stream transports the material which slope processes bring to the bottom of the valley (see Chapter 3).

The size of a river basin and the river which drains it vary greatly. For example, the Dunsop Basin (Fig. 4.8) contains rivers which drain areas as small as 0.06 km² (6×10^{-2} km²), while the world's largest rivers, such as the Amazon which drains an area of 5.776×10^6 km², are over 100 million times as large. In order to compare basins of different sizes, a classification system known as *stream ordering* has been developed.

(i) Stream order

The most widely used method of 'ordering' streams is that developed by A.N. Strahler. All the marked river channels which drain through one outlet can be traced from a large scale map, preferably 1:25 000 or larger. All those which have no tributaries and flow from a source are termed *first order* streams; where two first orders meet they form a *second order* stream; where two second order streams meet they form a *third order* stream, and so on (Fig. 4.8). It is useful to have different symbols to denote streams of each order. The system takes no account of distance and all fourth order basins are regarded as similar.

Fig. 4.8 Drainage basin of the River Dunsop, Forest of Bowland, Lancashire

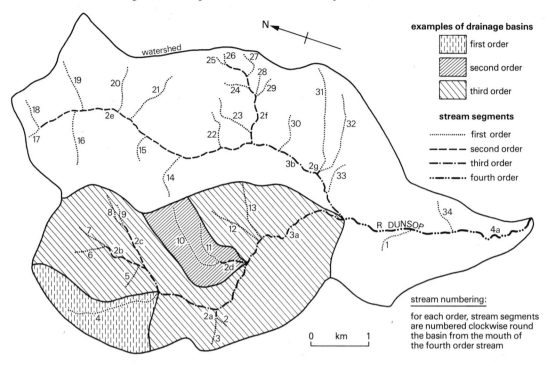

(ii) Stream number and order

If the number of streams (frequency) of each order is counted, it is easy to see that there are far more first than second order. If the frequency of each order is plotted on semi-log graph paper, the points can be linked with a straight line, or very nearly so (Fig. 4.9a).

This relationship between number of streams and order has been called *Horton's Law of Stream Numbers*, which can be stated simply as follows: there is a constant geometric relationship (ratio) between the number of streams of

Fig. 4.9 Morphometric relationships in the Dunsop basin

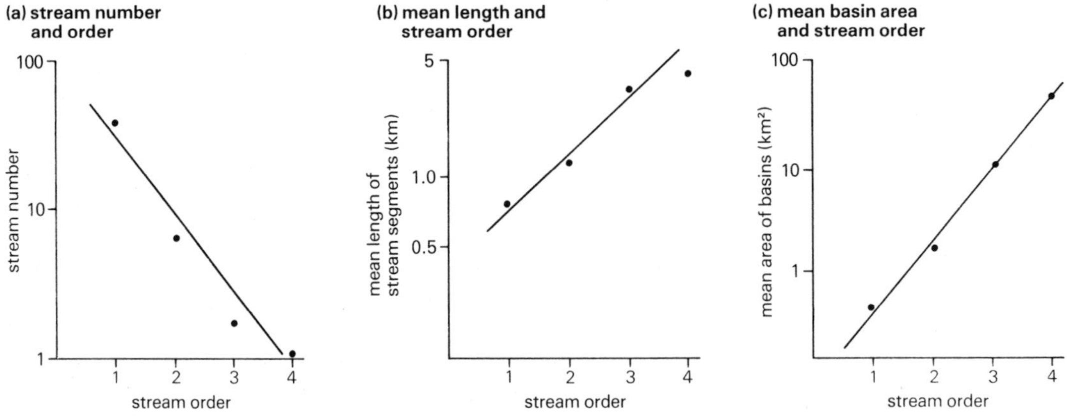

(a) stream number and order

(b) mean length and stream order

(c) mean basin area and stream order

one order and the next highest order. This can be expressed as a ratio called the *bifurcation ratio*, which indicates how many streams of one order, on average, are required to produce a stream of the next highest order.

To take an example, in the Dunsop Basin the number of first order streams (34) divided by the number of second order streams (7) produces a bifurcation ratio of 4.86:

$$\text{Bifurcation ratio of first order streams} = \frac{34}{7} = 4.86$$

The bifurcation ratio of large basins is the average of the bifurcation ratio of the stream orders within it. Table 4.1 shows the bifurcation ratio for the whole of the Dunsop Basin. The bifurcation ratio for most natural stream systems lies between 3 and 5.

Table 4.1 Bifurcation ratios for the River Dunsop

Stream order	Number	Bifurcation ratio
1	34	4.86
2	7	3.5
3	2	2.0
4	1	
Total = 10.36		

Average bifurcation ratio: $x = 3.45$

(iii) Stream length and order

If the length of each stream segment is measured and the average length of stream in each order is plotted on semi-log graph paper, it can be seen that there is again a fairly constant relationship between the two (Fig. 4.9b). The higher order basins generally have longer rivers.

(b) Areal properties

The measurements so far have all been linear, i.e. in one dimension. By introducing a second dimension, the areal properties of the basin can be measured. Using contour patterns, it is possible to delimit the area of the basin which contributes water to each stream segment (page 78). The watershed of the first order streams can be traced from where the stream joins the higher order stream along the hillcrests to pass upslope of the source and return to the junction. This line separates slopes which feed water towards the stream from those which drain into other streams. If the area of each first order basin is found in this way, it is possible to calculate the mean area of the basins. If the watersheds of the second order basins are traced in the same way, it will be seen that there are areas between the first order basins which drain directly into second order streams. These are called *inter-basin areas* and they mean that the area of the second order basin is not the sum of the first order basins within it. Plotting the mean area of basins of each order on semi-log graph paper shows that there is a constant relationship between the mean area of basins of successive orders (Fig. 4.9c).

(i) Density

The average length of channel per unit area of the drainage basin is called the *drainage density*. In the Dunsop Basin (Table 4.3):

$$\text{Drainage density} = \frac{\Sigma L}{\Sigma A} = \frac{43.21}{29.45} = 1.467 \text{ km/km}^2$$

where ΣL is the total length of all streams and ΣA is the total area of the whole basin.

This value indicates how frequently streams occur on the land surface. In upland areas with impermeable rocks, high rainfall totals and steep slopes, drainage densities are high. Conversely, low values of drainage density are found where slopes are gentle, rainfall low and the bedrock permeable. The density of vegetation in an area influences the drainage density by binding the surface layer, thus preventing overland flow from concentrating along definite lines, and from eroding small rills which might become stream channels. The vegetation slows down the rate of overland flow and in effect stores some of the water for short periods of time.

Table 4.2 illustrates the effect of climate and lithology on drainage density. In south-east England chalk areas have a high permeability, infiltration rates are high and, as a consequence, overland flow rates are limited. Active stream channels are relatively infrequent and therefore drainage density is low. Groundwater flow in this situation is relatively important. In contrast, the Badlands of Arizona have little vegetation and low permeability. Rainfall tends to occur in heavy downpours which produce large volumes of overland flow during and immediately after the storm. Channels are closely spaced, though often dry, and values of drainage density are high.

Table 4.2 Drainage densities and geology

Density (km/km²)	Texture description	Geology	Locality
2.7–3.5	Very coarse	Chalk Downs	S.E. England
3–8	Coarse	Carboniferous Limestone	W. Yorkshire
15–25	Medium	Weathered Igneous	California coast
25–40	Fine	Ordovician Lavas	N. Wales
200–900	Ultra-fine	Loess Badlands	Arizona and New Jersey

(ii) Shape

The shape of the basin is very important. Very long basins will take longer to achieve a throughflow of water from a rainstorm. The most efficient basin would be one in which the watershed is circular and all the water disappears down a hole in the middle. This would minimise the channel length in the basin. Two simple measures have been devised to assess shape:

Basin circularity compares the area of the basin to the area of a circle of the same circumference:

$$P = \pi D$$

$$\therefore D = \frac{P}{\pi}$$

$$A_0 = \pi \left(\frac{D}{2}\right)^2$$

$$\text{Basin circularity} = \frac{\Sigma A}{A_0}$$

where : P is length of basin perimeter of watershed

D is diameter of a circle equivalent in circumference to P

A_0 is area of circle of diameter D

ΣA is total area of basin.

This tells us how near the watershed of the basin lies to a circle. The basin circularity of the Dunsop Basin is:

$$P = 25.2$$

$$\Sigma A = 29.45$$

$$\therefore \quad D = \frac{25.2}{\pi} = 8.02$$

$$A_0 = \pi \left(\frac{8.02}{2}\right)^2 = 50.52$$

$$\text{Basin circularity} = \frac{29.45}{50.52} = 0.582.$$

Basin elongation compares the longest dimension of the basin to the diameter of a circle of the same area as the basin:

$$\Sigma A = \pi r^2$$

$$\therefore \quad r = \sqrt{\frac{\Sigma A}{\pi}}$$

$$\text{Basin elongation} = \frac{DL}{D}.$$

This indicates how nearly circular the *area* of the basin is; the nearer 1, the greater the correspondence to a circle. The basin elongation of the Dunsop Basin is:

$$\Sigma A = 29.45$$

$$DL = 9.25$$

$$r = \sqrt{\frac{29.45}{\pi}} = 3.06$$

$$\text{Basin elongation} = \frac{9.25}{2 \times 3.06}$$

$$= 1.51.$$

(c) Relief properties

So far all the features we have considered have been assumed to lie on a plane surface. The third dimension introduces the concept of relief. By measuring the vertical fall from the head of each stream segment to the point where it joins the higher order stream and dividing the total by the number of streams of that order, it is possible to obtain the average vertical fall. If this is plotted against the average stream length of the order, the *average gradient* is obtained. The average gradient of Dunsop Basin is:

Total fall of first order streams = 3575 m

Total length of first order streams = 24.49 km

$$\text{Average gradient} = \frac{3575}{24.49 \times 1000}$$
$$= 146 \text{ m/km}.$$

The average gradients of streams of each order when linked together produce an average long profile of the basin as in Fig. 4.10, which shows the long profile of the Dunsop Basin. Generally these long profiles illustrate that the lower order tributaries are steeper than those of the higher orders. This relationship has been observed in geomorphology for a long time and is dealt with more fully on page 119.

Fig. 4.10 Long profile of the River Dunsop constructed from averaged data

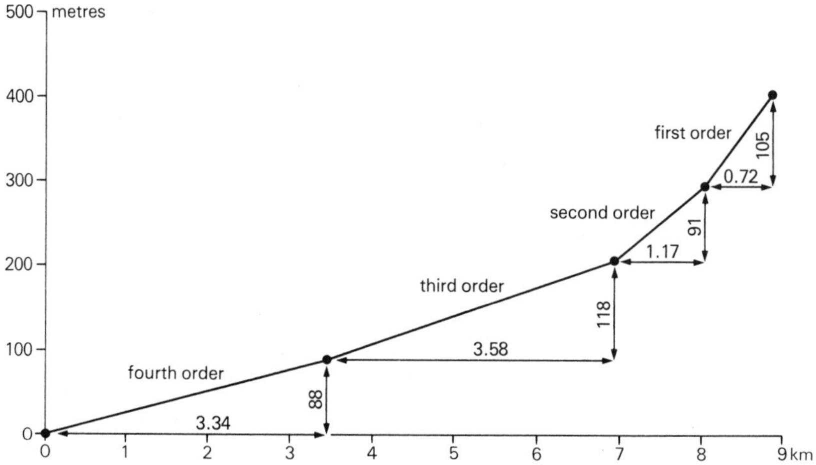

(d) Network developments and time

The relationships between the linear, area and relief properties of drainage basins are in themselves merely descriptive. They enable us to compare areas objectively. They are measures of part of the landscape. We need to go further and interpret the meaning of these figures.

An illustration is provided by a study of drainage densities on boulder-clay deposits of known ages which was carried out near Des Moines, Iowa. Each ice advance during the Pleistocene glaciation deposited a layer of boulder-clay (page 188). Successive advances did not always cover the whole of the area of the previous advance. Some areas have a surface layer of boulder-clay which is very old, while the surface in other areas might be in a boulder-clay deposited as recently as 13 000 years ago. The drainage patterns developed on two such areas are shown in Fig. 4.11. The older area has a much more intricate pattern and its drainage density, measured using the method described on page 86, is much higher.

Fig. 4.11 Development of drainage patterns over time on Des Moines lobe, Iowa, USA (After Ruhe)

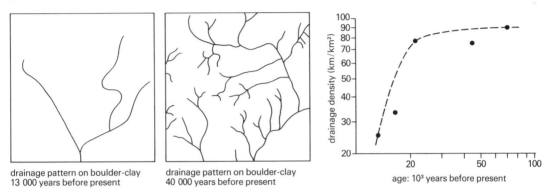

drainage pattern on boulder-clay 13 000 years before present

drainage pattern on boulder-clay 40 000 years before present

age: 10^3 years before present

From a number of determinations of drainage density on boulder-clays of different ages, the graph in Fig. 4.11 was constructed. It shows that drainage density increases rapidly at first and then levels off at a constant value. This value is a complex function of rainfall, slopes, vegetation and the depth and permeability of the regolith.

(e) *Conclusions*

At the end of this section, a word of caution is perhaps necessary. Some of the work involved in the collection of data for these morphological analyses of stream networks is very time-consuming. Third and fourth order basins in humid areas are of manageable dimensions for one person. Higher order basins frequently cover very large areas and consequently require many hours to complete. Bear in mind that the Mississippi Basin is a tenth order basin, which is obviously beyond the bounds of practicability for one person.

There are several possibilities for 'automating' the measurements:

1. Measuring the lengths of stream segments can be considerably speeded up by using a small map measuring wheel. These can be used quickly enough to enable measurements to be repeated and hence the accuracy increased.
2. The tedious method of measuring area by counting squares on graph paper can be eliminated by using a planimeter or, less expensively, by cutting the areas from a good quality paper and weighing them on an accurate balance. These weights are easily converted to area if the weight of paper is known for a unit area at map scale.

Table 4.3 Morphometric data for the Dunsop basin

Name of stream	Dunsop	Grid reference of mouth of basin	SD 662500	

Order	Number	Length (km)	Area (km²)	Fall (m)
1	1	0.86	0.50	78
	2	0.29	0.12	30
	3	0.40	0.12	61
	4	1.73	1.29	167
	5	0.36	0.23	70
	6	0.56	0.42	90
	7	0.57	0.36	90
	8	0.53	0.33	41
	9	0.26	0.06	21
	10	1.37	0.77	148
	11	0.81	0.22	124
	12	1.50	0.45	179
	13	1.51	0.66	94
	14	0.57	0.60	122
	15	0.52	0.47	92
	16	0.98	0.55	138
	17	0.22	0.13	18
	18	0.55	0.17	60
	19	0.82	0.45	181
	20	0.84	0.56	130
	21	1.13	0.85	137
	22	0.78	0.25	170
	23	0.95	0.26	195
	24	0.39	0.29	102
	25	0.11	0.06	8
	26	0.12	0.06	10
	27	0.51	0.12	59
	28	0.17	0.12	114
	29	0.34	0.15	138
	30	0.56	0.35	144
	31	1.71	0.58	209
	32	1.34	0.56	203
	33	0.51	0.44	70
	34	0.62	0.22	82
	$N_1 = 34$	$L_1 = 24.49$	$A_1 = 12.77$	$F_1 = 3575$
2	a	0.21	0.29	41
	b	0.85	1.60	61
	c	0.80	1.33	64
	d	0.42	1.21	49
	e	4.25	8.09	170
	f	1.57	1.64	233
	g	0.11	1.28	21
	$N_2 = 7$	$L_2 = 8.21$	$A_2 = 15.44$	$F_2 = 639$
3	a	4.90	11.15	182
	b	2.27	14.47	53

Order	Number	Length (km)	Area (km²)	Fall (m)
	$N_3 = 2$	$L_3 = 7.17$	$A_3 = 25.62$	$F_3 = 235$
4	a	3.34	29.45	88
	$N_4 = 1$	$L_4 = 3.34$	$A_4 = 29.45$	$F_4 = 88$

Name of stream ....Dunsop.......... Grid reference of mouth of basin ....SD 662500..........

Basin perimeter = 25.2 km
N = number of streams of each order;   L = length of streams of each order;
A = area of basin of each order;   F = fall of streams of each order.

3. The problem of calculation is eased if the data is recorded systematically. A suggested method is shown in Table 4.3. This table is also a summary of the work included in this section.

It is not necessary to complete all the measurements for a particular stream system, nor would it be very wise in view of the amount of work required. It is possible to make one measurement, such as a drainage density, in a number of basins in order to contrast them, as illustrated on page 89. The relationships between streams of different orders are simple to understand and are surprisingly regular.

The usefulness of methods like this is that:

(a) they may enable us to predict, with a known degree of error, properties of higher or lower order drainage basins;

(b) they provide an objective way of examining these units of landform and of comparing one area with another. In the next section we build further upon this basis.

ASSIGNMENTS

1. (a) *Figure 4.12 shows a small fourth order drainage basin. Using the methods described on page 85, calculate:*
   (i) *the bifurcation ratio;*
   (ii) *the drainage density.*

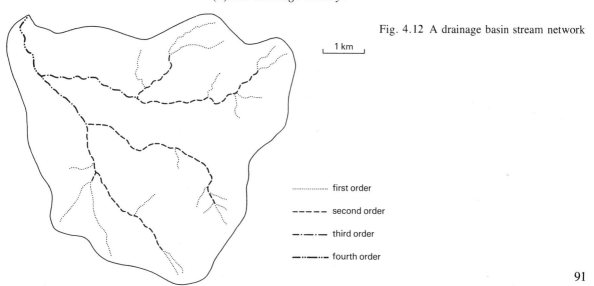

Fig. 4.12 A drainage basin stream network

1 km

.... first order

– – – – – second order

–·–·–·– third order

–··–··– fourth order

(b) *What conclusions can you draw about this basin in comparison with the River Dunsop (page 84)?*

(c) *Describe the methods you might use to compare the shape of this basin with that of the Dunsop.*

(d) *Construct a graph to illustrate the relationship between the stream length and order in this basin.*

2. *The drainage patterns for two areas each 5 km by 5 km are shown in Fig. 4.13. The highest and lowest points are shown on the maps and a brief outline of their rainfall and geology is given beneath the diagram.*

(a) *Calculate the drainage density of each area.*

(b) *Explain how the differences in the density of the drainage pattern may have originated with reference to slopes, rainfall and geology.*

Fig. 4.13 Drainage networks in the Cotswolds and Northumbria

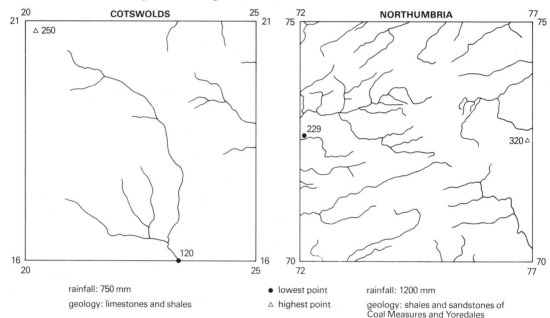

rainfall: 750 mm

geology: limestones and shales

● lowest point

△ highest point

rainfall: 1200 mm

geology: shales and sandstones of Coal Measures and Yoredales

C. River Channels

1. Energy variation in rivers

A still body of water at any point above sea level has a certain amount of stored energy as a result of its position. This is *potential energy* and it is available to do *work* in the river channel. Its quantity is proportional to the mass of the water body and to the height through which it has to fall to return to the sea. The energy originates from the sun which evaporates water from the sea enabling it to be deposited at a higher level as precipitation over the land (page 77).

The *kinetic energy* of a river is caused by its movement and is derived from the potential energy. Between the two there is some 'loss' as energy is needed to overcome the internal friction of the water, or its viscosity, and the friction with the bed of the channel. The kinetic energy represented by the velocity of the water in the channel does work in eroding and transporting sediment. Some energy is used in creating sound and heat but the quantity is very small.

If the channel gradient is steep, the change from potential to kinetic energy is rapid and the velocity of the river is high. Conversely, on gentle gradients the velocity is lower. The amount of work done by a river depends not only on its velocity but also on the mass of water, or its volume. The volume passing a point in a given time, usually one second, is the *discharge*. The greater the discharge, the larger the total energy of the river. If follows that large channels have a greater total energy than small ones.

In summary of this section, we can say that the kinetic energy of the river, represented by its velocity, is determined by the internal friction of the water, the bed friction, the slope of the channel, the discharge and the size of the channel. To assess the importance of these variables we must first understand the way in which a river flows.

2. Types of flow

When a water particle flows along an open channel it does not follow a direct straight line or even a smoothly curving path or trajectory. It moves vertically and laterally in addition to its overall downstream direction. The actual velocity of this particle is much greater than the mean velocity in a downstream direction, which might be measured over a short time by a method such as described on page 115. This type of flow is called *turbulent flow*, and predominates in most natural river channels (Fig. 4.14b). It accounts for their efficiency in eroding and transporting sediment (see page 107).

A second type of flow, *laminar flow*, (Fig. 4.14a) is recognised by the nearly linear trajectories of water particles. It is found at lower flow velocities and is of very limited occurrence in natural river channels.

The shapes of the vertical velocity profiles of these types of flow are different in that the profile of the laminar flow is a smooth *parabola* curving up from the stream bed, and that of turbulent flow is, in comparison, sharply angular in the zone immediately above the stream bed. This reflects the change from laminar flow at the bottom of the profile, where the water apparently slides or shears over itself in layers like a pack of cards pushed from the top, to turbulent flow, where velocity distribution is more random and the velocity profile more uniform. The narrow zone of laminar flow in the profile is thought to be present around most objects on the river bed. In these areas velocity is low because of the friction between the static bed and the moving water. The way in which the water overcomes this is to shear over itself producing a laminar flow pattern, exerting a shear force on the obstacle or bed.

In plan, the variation in the velocity is affected by the same factors as the vertical velocity profile. The friction is greatest at the edges of the channel, which are therefore the zones of slowest flow. The fastest water is usually found in the centre of the channel (Fig. 4.14c). This simple flow pattern is disrupted if the channel shape is more complex. A large boulder in the bed may effectively divide the channel into two, producing two velocity peaks in the plan.

Fig. 4.14 Velocity profiles and channel measurements

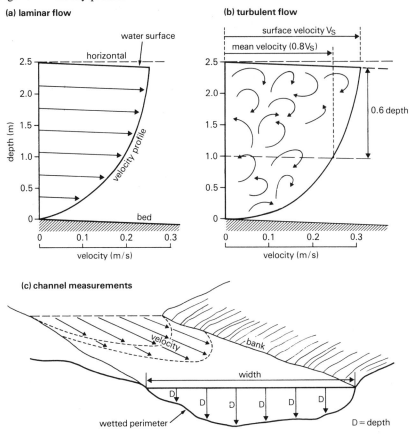

(a) laminar flow

(b) turbulent flow

(c) channel measurements

3. Velocity and discharge

The velocity varies vertically and laterally across the river, so that to speak simply of 'velocity' can be confusing. To be more exact, we refer to the *mean* or *average velocity* which is the average of a large number of velocity measurements made in a cross-section of the river. This is often impractical, since the river might be so shallow as to allow only one measurement to be made at each position across the section, or there may be a time limit to the operation which restricts the number of readings. Frequently, velocity is measured only at the surface, and Fig. 4.14 shows that this is the fastest point in the vertical profile. Such readings of velocity are generally multiplied by a factor of 0.8 to give an approximation to the mean velocity. Where a single reading is taken at depth at each point across the section, the depth is normally 0.6 of the total depth at that point, measured down from the water surface. A number of methods of measuring the velocity of a river are explained at the end of this chapter.

The volume of water passing a point in a given time, *the discharge*, is dependent on the velocity of the river and the size of the channel at that point. Apart from the velocity, all we need to know is the cross-sectional area of the river. The discharge is then calculated by multiplying the area in m² by the velocity in m/s to give the discharge in m³/s. For example, if the river has an average velocity of 0.2 m/s and a cross-sectional area of 10 m², the discharge is (0.2 × 10) m³/s or 2 m³/s. Usually the equation is written:

$$Q = A \times V$$

where Q represents discharge, A the cross-sectional area and V the mean velocity. Applying this to Fig. 4.15a:

$$A = 11.32 \text{ m}^2 \qquad V = 0.153 \text{ m/s}$$
$$Q = (11.32 \times 0.153) \text{ m}^3/\text{s}$$
$$= 1.73 \text{ m}^3/\text{s}$$

Fig. 4.15 River cross profiles

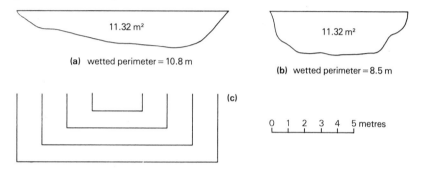

(a) wetted perimeter = 10.8 m

(b) wetted perimeter = 8.5 m

(c)

0 1 2 3 4 5 metres

The discharge of course varies from times of flood to times of drought and as the level of the river rises and falls, so will the cross-sectional area and the velocity.

Measuring the cross-section, in theory at least, is relatively simple. The line of the section is the shortest distance approximately at right angles to the flow across the river. The length along this line from the water's edge to the opposite edge is the *width* (Fig. 4.14). The *depth* is the vertical distance from the water surface to the bed of the channel. If the depth is measured every metre or half metre across the width, a cross-section like those in Fig. 4.15 can be constructed. The *cross-sectional area* can be measured from such a diagram if the vertical and horizontal scales are the same. Alternatively, the mean of the depth measurements multiplied by the width gives a comparable result. From the diagram of the cross-section it is possible to measure the length of the bed in contact with the water. This is shown in Fig. 4.15 and is known as the *wetted perimeter*.

An alternative method for measuring discharge in small streams is the 'V' notch weir shown in Plate 4.1. A 90°V cut in plywood forms part of a temporary dam across the stream. The edge of the V should be as smooth as possible and the 'dam' can be extended by using sheets of polythene to make it watertight. The V acts as a spout. If the flow rate is very low, the water can be caught in a bucket and the discharge calculated. Where the discharge is greater, it can be calculated from:

$$Q = 1.336 \ h^{2.48}$$

Plate 4.1 A V-notch weir (Author's photograph)

where h is the height of the level water surface above the angle of the V. This height must be measured where the surface is unaffected by the 'downdraw' through the notch. If the height above the V is 75 mm, the discharge would be:

$$Q = 1.336 \times 0.075^{2.48}$$
$$= 0.00217 \text{ m}^3/\text{s}$$

or just over two litres per second.

4. Friction of water with the bed

The effect of the bed on flow is determined by how much of the water comes into contact with it. This depends on the length of the wetted perimeter and the cross-sectional area of the channel.

The relationship between the wetted perimeter and the cross-sectional area is termed the *hydraulic radius*. It is calculated from the equation:

$$R = \frac{A}{WP}$$

where R is the hydraulic radius, A is the cross-sectional area and WP the wetted perimeter. In Fig. 4.15:

$$A = 11.32 \text{ m}^2 \qquad WP = 10.8 \text{ m}$$
$$R = \frac{11.32}{10.8} \text{ m}$$
$$R = 1.05 \text{ m}.$$

This value indicates that if the bed or wetted perimeter were stretched out flat and the water in the cross-section were heaped upon it, it would form a layer 1.05 m deep. In effect, each metre of the wetted perimeter affects 1.05 m² of the water in the cross-section. If the value of the hydraulic radius is large, a large area of water in the cross-section is affected by each metre of the bed. The frictional effect of the bed is therefore very limited. If the value is small, the frictional effect is large, as in a very shallow river.

The friction of the bed of the river itself varies with how 'rough' it is. If the bed is a smooth surface of silt, it has a much lower frictional effect than if it is composed of gravel and large boulders. This *bed roughness* is very difficult to measure directly but it can be placed on a relative scale by comparing it with other stream channels and describing it as 'very rough' or 'moderately smooth' etc. Greater roughness will tend to slow down the river, while a smooth channel bed will produce a higher velocity.

5. Manning's equation

The total effect of these varied factors of internal friction, bed roughness, channel slope, size and shape and the discharge on the velocity of the river is difficult to assess when they are all considered together. However, in 1889 Manning, by experiment, devised a formula which illustrates how we can think about these variables. Using the same symbols as before:

$$Q = A \times \frac{R^{\frac{2}{3}} \times S^{\frac{1}{2}}}{n}$$

where n is Manning's coefficient of bed roughness, some values of which are given in Table 4.4. Notice that the higher the value the rougher the bed will be.

Table 4.4 Manning's coefficient of bed roughness

Surface	Manning's n
Very smooth e.g. glass	0.010
Concrete	0.013
Unlined earth drainage channels	0.017
Winding natural channels	0.025
Mountain streams with rocky beds and rivers with variable section	0.040–0.050
Alluvial channels with small ripples	0.014–0.024
Alluvial channels with large dunes	0.020–0.035

The only difference between this formula and the discharge equation on page 95 is that the velocity (V) has been replaced by values of R (hydraulic radius), S (slope) and Manning's n. The slope is measured as a fraction, the vertical fall of the water surface in the vicinity of the section divided by the distance horizontally in which that fall takes place. For example, if the river falls 1 m in 1000 m, the slope is 0.001. If we take cross-section (b) in Fig. 4.15, the area (A) is 11.32 m^2, the wetted perimeter (WP) is 8.5 m and the hydraulic radius (R) is:

$$R = \frac{A}{WP} = \frac{11.32}{8.5} = 1.33 \text{ m.}$$

If the estimated value of Manning's coefficient is 0.025, indicating that the channel is a 'winding, natural' one, probably with a sandy bed, the discharge is calculated as follows:

$$Q = 11.32 \times \frac{(1.33)^{\frac{2}{3}} \times (0.001)^{\frac{1}{2}}}{0.025}$$
$$= 11.32 \times \frac{1.21 \times 0.03162}{0.025}$$
$$= 17.32 \text{ m}^3/\text{s}$$

From this it is relatively easy to see that if the bed roughness increases, i.e. n is larger, the velocity and hence discharge are reduced, and if the slope or hydraulic radius is increased, the velocity and discharge will also increase.

Manning's formula is useful in estimating the discharge in flood conditions. The height of the water can be assessed from debris stranded in trees or high on the bank. All that need be measured is the cross-section (page 95) and the slope. Manning's n value can be estimated from Table 4.4. Floods are dealt with more fully on page 108.

6. Results of research

In 1953 Leopold and Maddock published the results of their research into the 'hydraulic geometry' of stream channels. They found that for many streams, both in 'normal' and 'flood' flow, stream width, depth and velocity increase as simple power functions of discharge. From records of streams and rivers in central and south-western USA, they obtained the direct relationships:

Fig. 4.16 Downstream changes in width, depth and velocity (After Leopold *et al.*)

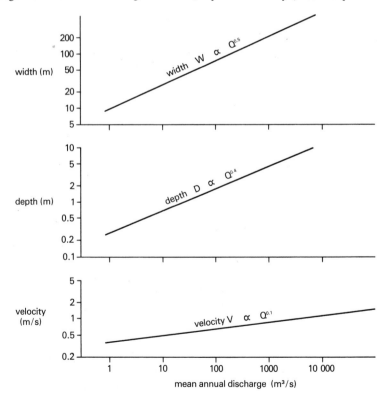

$W \propto Q^{0.5}$

where W is width and Q is discharge

i.e. the width of the channel increases as the square root of discharge;

$D \propto Q^{0.4}$

where D is mean depth

i.e. the depth of the channel increases nearly as the square root of discharge;

$V \propto Q^{0.1}$

where V is mean velocity

i.e. the velocity increases as the tenth root of discharge.

These relationships are shown graphically in Fig. 4.16. As volume of water flow and hence discharge increase downstream, it follows that width, mean depth and mean velocity all increase with increasing distance from the source. It was an accepted fact that rivers became deeper and wider further downstream but it came as a surprise to many to learn that the average annual current velocity also *increases* downstream. It is now thought that a river flows more efficiently when its depth increases. This more than compensates for the reducing angle of channel slope and thus a slight increase in velocity occurs (Fig. 4.16).

7. Variations in discharge

The variations in the discharge of a river are the result of changes in precipitation, evapotranspiration and of storage within the basin, as outlined on page 80. Rivers in Britain clearly will rise in level and so in discharge after periods of rain. In this section we are going to examine how an individual rainstorm affects the discharge and how the discharge varies over the year.

(a) The storm hydrograph

A rainstorm over a river basin causes the discharge to change in a way similar to that shown in Fig. 4.17. Over a period of time the river rises to a *peak discharge*, the *rising limb*, and then falls to a level similar to that before the storm, the *recession limb*. This starting and finishing level is called *base flow* and it represents the part of river flow produced by groundwater seeping slowly into the channel bed. The low flow velocities through rock and regolith produce a steady input of water which varies relatively little in comparison with the discharge of the whole river. On the hydrograph the base flow is determined by joining the lowest point of the rising limb to the point of inflection on the recession limb as in Fig. 4.17.

After the rainfall has started, the first water to reach the gauging station or measuring point is that which falls directly into the streams within the basin. This is referred to as *channel catch* and causes the initial rise of the stream. During the early stages of the storm, infiltration into the soil is high, and a large part of the precipitation enters the regolith to contribute to base flow. If the storm continues, overland flow (page 80) begins to occur and water runs over the slopes of the basin and, via storm channels, enters the permanent

Fig. 4.17 The storm hydrograph

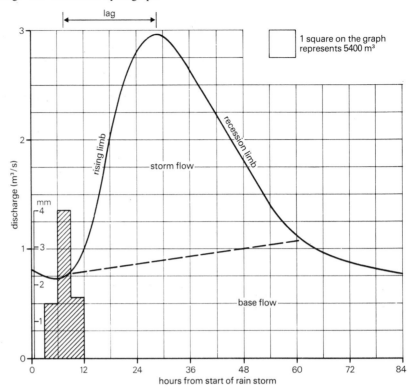

streams. The rising limb is mainly the product of this overland flow. Its shape depends on the intensity and duration of the rainstorm itself and the state of saturation of the regolith at the start of the storm. Prolonged light rain on a dry soil will give a lower, flatter hydrograph than a short heavy shower on a saturated soil, even though both storms produce the same total rainfall.

The time taken for water to flow from the point where it hits the ground to the gauging station results in a delay between the rainfall and the peak discharge, which is called the *lag period*. It is usually measured between the rainfall peak and the discharge peak and can vary from a few minutes in the case of a very small basin, to many days for a very large one.

The *recession limb* of the hydrograph is determined by the nature of the basin rather than by that of the storm, so that for any particular basin the recession limb is much more predictable than the rising limb. This part of the hydrograph represents water draining through the soil and over the surface as the storm dies away, and from the channels (channel storage) as the level of water in them falls. When all the water from the storm has passed, the stream returns to base flow. If there is no rainfall, base flow will very slowly decline as the level of the water table falls. If this level falls lower than the channel bed, the river will cease to flow and will sink into the bed if the rock is permeable. The river will resume its flow when the groundwater is recharged by the next storm. Such conditions are common in small streams and in arid and semi-arid areas where rain is infrequent. Areas of highly permeable rocks or low water tables often have dry valleys (page 126).

(b) The size of river basins

The storm hydrographs resulting from the same storm in two different sized basins are shown in Fig. 4.18. The smaller basin is about 24 km² and has a lag period of 1 hour, while the larger basin is 140 km² and has a 4 hour lag period. This is the result of the longer distances over which the water has to flow in the larger basin. There are other factors which affect the lag period. Steep slopes accelerate run-off, while a dense vegetation cover retards it. Further effects are examined on page 110.

Fig. 4.18 The relationship between discharge and area in two river basins

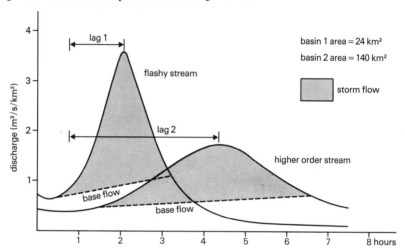

In Fig. 4.18 the 'y' axis does not show the actual discharge in the basin but the discharge per square kilometre. This enables the two basins to be compared more easily as it negates the effect of their difference in size. Both basins yield approximately the same total volume of water per square kilometre as a result of this storm, but in the small basin this passes the gauging point quickly, giving about 3 hours of *storm flow*, while in the larger basin the storm flow period is 5 hours. The larger basin nevertheless has a much greater total discharge since it gathers water over an area of 140 km².

Streams like the smaller one in Fig. 4.18 are described as *flashy*, since they rise and fall quickly and occasionally produce *flash floods*. Their lag periods are short and their peak discharges high in relation to base flow. In larger rivers the lag is longer and river levels tend to be more constant.

(c) The annual hydrograph and river regime

The continuous recording of river discharge is carried out in Britain by the Regional Water Authorities. This enables the annual hydrograph to be drawn fairly easily for most major rivers. Figure 4.19a shows the hydrograph for the River Coquet in Northumbria. The dashed line indicates the level of base flow, while above it the peaks of the storm flow indicate periods of heavy rain in the basin. In 1969 there appear to have been more individually recognisable storms than in the preceding year, while the months of July, August, September and October saw very low flows in the river and base flow fell to an almost constant low at about 0.1 m³/s. This basin has an average rainfall of about 1200 mm/a and the equivalent of 755 mm passes the gauging point as run-off. This means that approximately 37 per cent of the rainfall leaves the basin by other means, mainly as evapotranspiration.

The mean daily or monthly discharge measured over a period of years shows a pattern. In Britain the lowest discharges usually occur in late summer when soil moisture and groundwater flow are at their lowest. This 'average' situation is referred to as the *regime* of the river. It is analogous to 'climate'; the annual hydrograph is the 'weather' of the river. The regime of the River Tees is shown in Fig. 4.19b with the discharge plotted on a log scale. Highest flows are

Fig. 4.19 Discharge of the River Coquet and regime of the River Tees (Crown Copyright Reserved, reproduced from *The Surface Water Yearbook* with permission of the Controller of Her Majesty's Stationery Office)

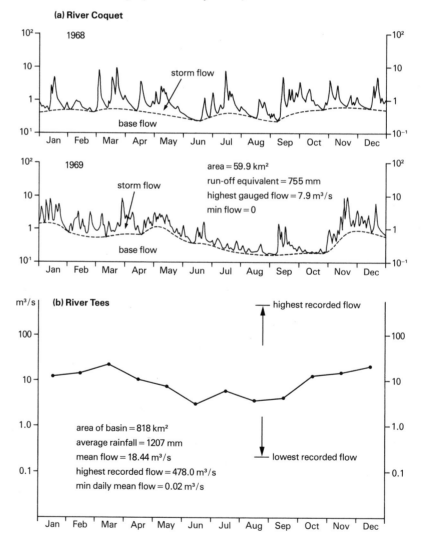

normally recorded in spring when evapotranspiration is low, and snowmelt from the high Pennine moorlands releases some stored water. The absolute maximum and minimum flows are also shown to illustrate the variation which the averaged values hide. On a world scale the regime of rivers varies widely as a result of variations in climate. Some rivers cease to flow at certain times of the year, for example those in the Canadian Arctic, while others such as the Amazon vary little in relation to their mean discharge.

ASSIGNMENTS

1. *A river has a mean depth of 0.6 m and a width of 6.2 m. Its wetted perimeter measures 9.3 m. The channel at this point falls 0.3 m over a horizontal distance of 100 m. The discharge measured a short distance below this section is 2.77 m^3/s.*
   (a) Calculate the hydraulic radius of this channel. (Page 96)
   (b) What is the value of Manning's coefficient of bed roughness for this section of the river? (Page 97)
   (c) What type of bed does the n value you have calculated indicate at this point? (Page 97)
   (d) What is the average velocity of flow?
2. *(a) From the curve in Fig. 4.17, calculate the volume of water which passes the gauging station as storm flow. This can be done by measuring the area under the curve by counting the squares on the graph paper.*
   (b) If the area of the basin is 15 km^2, what is the run-off in millimetres rainfall equivalent? Calculate the total rainfall from the rainfall data. Explain the differences between these figures.
3. *Draw a storm hydrograph using the figures in Table 4.5. Compare this hydrograph with that in Fig. 4.17. What can you infer about the two basins from the differences in the two hydrographs assuming that the basins have similar areas?*
4. *How can geology affect a storm hydrograph?*

Table 4.5 Storm discharge data

Time (hours after start of rain)	Discharge (m³/s)	Time (hours)	Discharge (m³/s)
0	0.54	36	0.81
0	0.87	42	0.68
4	1.14	48	0.60
6	1.52	54	0.57
8	1.79	60	0.55
10	1.97	66	0.53
12	2.30	72	0.50
18	1.87	78	0.49
24	1.25	84	0.50
30	0.94		

D. Erosion and Transportation

1. Methods of erosion

Water which is draining towards the outlet of a river basin, i.e. run-off, has a kinetic energy which is used to do work. Geomorphologically, the most important outcome of this work is the wearing away and removal of material from the wetted perimeter of the channel. This is referred to as *erosion*, and the eroded material is transported along the channel. Erosion may take place in solid bedrock, in unconsolidated alluvial material or in slope deposits. In this last case the river acts as the transport agent for the 'slope conveyor' (page 67). All the material transported by the river is known as the *load* and the ways in which it is transported are examined later in this section.

The ways in which erosion is accomplished can be divided into four:

(i) *Hydraulic action* moves material in the channel through the impact of the moving water and its frictional drag on the particles lying on the bed. Usually it is capable of removing only unconsolidated sediment such as sand or fine gravel. The velocities attained in normal river flow have little effect on hard rocks. Hydraulic action is easy to demonstrate by running a hose into a tank of water containing sand. As the velocity of flow is increased the turbulence lifts more and more sand grains from the bottom. River banks are often undermined by hydraulic action and eventually they collapse into the river.

(ii) *Abrasion* occurs when rock particles already being moved by the river strike or are dragged along the rock bed. Their action is like that of a hammer chipping the rock or like a file smoothing down the surface and producing very small particles which are easily transported. If the particles in the load are large, erosion is more rapid. Small particles tend to smooth or polish the surface.

Since the river is using rocks as tools, abrasion is possibly the most effective method of erosion and is normally responsible for most of the downcutting in a river channel. If there is very little load in solid form, abrasion is ineffective, just as a smooth file would be. Rivers at base flow levels carry very little load in comparison to their 'brown' sediment-laden state during storm flow conditions. More abrasion occurs during these short violent periods than in months of low flow conditions.

Potholes in the beds of rivers are a common feature of rapid abrasion. They are cylindrical holes 'drilled' into the rock by turbulent high velocity flow. Vertical eddies may be strong enough to rotate a small pebble which grinds a depression in the rock. Plate 4.2 shows a pothole at low water level. The 'grinder' pebbles can be seen at the bottom waiting to be activated by the next storm to raise the river into flood. Potholes vary in size from a few centimetres to several metres in diameter. Very large ones are found beneath glaciers where water velocities are very high. Adjacent ones may join together creating sudden and considerable deepening of the channel, such as at the Strid near Bolton Abbey on the River Wharfe in Yorkshire.

(iii) *Solution* is described on page 25 in the chapter on Rocks and Weathering. It is most marked on carbonate rocks, such as limestones, where carbonic acid dissolves the rock and it is carried away in *solution*. Yet almost all rocks are

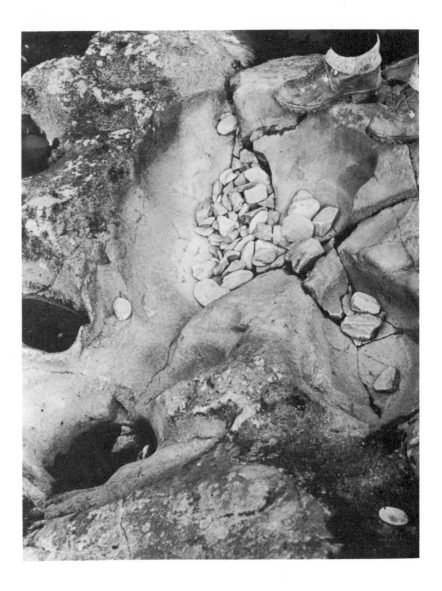

soluble to some extent, even silicates like quartz are found in solution, particularly in tropical rivers where a large proportion of the load consists of dissolved material.

(iv) *Attrition* is the reduction in size of the particles in transport as they strike one another or the bed of the channel. Thus as any particle moves downstream there is a progressive reduction in size. In addition the sharp edges and angles of the particle become rounded since they are more exposed. The upper reaches of a river therefore tend to contain large, angular sediment while the downstream parts have fine, rounded particles.

This statement, though normally true, must be treated as a generalisation. Angular particles of large size can often be found in the lower reaches of a river. The angularity and size of any particle depend on its lithology, the length of time it has been in the river and on the distance it has travelled. Some pebbles may be rounded by other agents, such as glaciers or waves, before entering the river. In assessing size or angularity of the material in a river bed, a mean value calculated from a sample of a number of the particles must be used. Ways of measuring these are explained on pages 116 and 117.

2. Methods of transportation

Once material has been added to the load of the river by erosion, it is transported in three ways:

(i) *Suspension*   Some particles, such as silt and clay, are small enough to be held up by turbulence within the water, and form the *suspended load*. The more turbulent the water, the larger the particles which can be transported in suspension.

(ii) *Traction and saltation*   Larger particles, such as sand and gravel, roll and slide (traction) or bounce (saltation) along the bed of the river under the hydraulic force of the moving water. Together these particles moving close to or along the bed of the river are referred to as the *bed load*.

(iii) *Solution*   The dissolving of rocks adds to the river's load and is known as the *dissolved load*.

3. Velocity and transportation

Under normal conditions of flow the total energy of the river is small in comparison with that of storm flow conditions. The river's velocity is increased as the discharge rises. The higher velocity makes the river competent to carry larger particles. The largest transported particle gives a measure of the *competence* of the river. The greater volume of water passing down the channel enables a greater volume of load to be carried and the *capacity* of the river is increased.

The relationship between particle size, erosion, transport and deposition, and stream velocity has been determined experimentally and is shown in Fig. 4.20. As an example, consider a particle of sand 0.2 mm in diameter. If the velocity of the flow rises to 300 mm/s, it will be eroded from the river bed, but when the velocity falls deposition will not occur until the speed of the water is about 10 mm/s. Transport requires much less energy than erosion.

From Fig. 4.20 it can be seen that once a particle is eroded from the channel bed it can be transported by a lower velocity and thus it can be carried quite a long way before the velocity decreases and causes the particle to be deposited. The graph also shows that the small particles of sizes less than 0.1 mm in diameter require disproportionately larger velocities to raise them from the channel bed. This is because of the cohesion of the particles towards each other. In addition, as they lie on the channel bed, they offer less resistance to water flow than larger particles and thus require a more energetic stream to *lift* them.

It is also fairly clear from Fig. 4.20 that under flood conditions, when greater velocities are experienced, hydraulic action, abrasion and attrition will all operate more effectively and thus the effectiveness of erosion will be significantly increased.

106

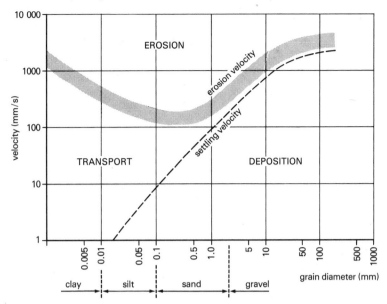

Fig. 4.20 Velocity and particle movement (After Hjulström)

Finally, as the velocity of flow determines the size and amount of material that can be transported, particles which constitute part of the bed load under low flow conditions in flood may be carried along as part of the suspended load. The increase in turbulence creates stronger vertical eddies which support the larger sized load. Many rivers in flood have a colouration due to the load. Frequently this is brown, but rivers issuing from glacier snouts are often a milky white or grey, while rivers from desert areas may be red.

ASSIGNMENTS

1. (a) *Using Fig. 4.20 what would be the minimum size of loose material with which you would floor a temporary crossing point in part of a river, assuming that the conditions in Assignment 1, page 103 prevailed?*
    (b) *Can you suggest reasons why this size of material might prove disadvantageous?*
    (c) *What would happen to discharge if you made the channel semicircular and lined it with rough concrete, keeping the same cross-sectional area and slope? (Page 97)*
2. *Explain which method of erosion you would expect to be dominant in a river channel composed of: (a) a hard sandstone; (b) a limestone; (c) a boulder-clay (till); (d) gravel.*
3. (a) *The cross-section of the river in Table 4.5 is 1.1 m² at 0 hours, 1.25 m² at 4 hours, 1.5 m² at 6 hours and 1.75 m² at 10 hours. Calcu-*

    late the velocity at each of these times using the formula $V = \dfrac{Q}{A}$

    (page 95). Graph the relationship between discharge and velocity.
    (b) *Using the graph constructed in (a), draw a curve to show the variation in velocity of the river during the passage of the storm flow.*

(c) *Using Fig. 4.20, at what time would: (a) fine sand, (b) clay, (c) gravel be eroded?*

(d) *How long would each of these sediments remain in transport?*

(e) *Explain how this storm discharge would affect the composition of the alluvium in the channel.*

E. Problems in River Management

Rivers are powerful agents of erosion and deposition. Fluvial processes are most active when the river is in flood and damage to land and property at its most severe. Expensive river constructions are needed to withstand specific flood conditions. In most cases their usefulness is confined to a few hours each year when the cost of their construction is justified.

1. Floods

Flooding occurs when the capacity of the channel to carry the discharge is exceeded. The channel is then said to be over the *bankfull stage* (Fig. 5.13).

(a) Recurrence intervals

Discharges which produce conditions over bankfull occur with widely differing frequency in different river basins. By examining records of discharge for previous years, it is possible to establish how frequently a particular discharge peak is likely to occur. This period is the *recurrence interval* and is calculated by first ranking the annual peak discharges with the highest first in the list. To find the recurrence interval of flow x m^3/s use the formula below:

$$\text{Recurrence interval of discharge } x \text{ m}^3/\text{s (in years)} = \frac{(\text{number of peaks in list}) + 1}{\text{ranked position of discharge } x}$$

A discharge of 18 m^3/s which was ranked twelfth in records extending over 40 years would have a recurrence interval of:

$$\frac{40 + 1}{12} = 3.42 \text{ years}$$

It is probable that a discharge of this magnitude would occur once in every 3.42 years. When we speak of the 20 year flood or the 100 year flood, we are not really referring to a time but to the frequency with which a particular flood height can be expected to return.

(b) Economic consequences of floods

Extremely high discharges occur with decreasing frequency in any particular drainage basin. However, they are frequently catastrophic. The River Dunsop (page 84) experienced such an event on 8 August 1967. One rain gauge, the only one to survive, recorded 116.8 mm in 1.5 hours. A nearby river rose 5.8 m in 45 minutes, and in the village of Wray three houses were destroyed and ten more severely damaged. Damage was estimated at £100 000.

One of the most damaging British floods occurred in the Devonshire village of Lynmouth on 15 August 1952. Possibly as much as 250 mm of rain fell in the 100 km² of the Lyn catchment in 24 hours, and a calculated discharge of 511 m³/s occurred at the height of the storm, a discharge only exceeded by the Thames once in every fifty years.

It was estimated that 100 000 tonnes of boulders were moved into the village and one of 7.5 tonnes was found in the basement of an hotel. A boulder dump of a previous flood some distance upstream of Lynmouth remained unmoved, indicating that higher discharges had occurred in the past. The cost of this flood was estimated to be in excess of £10 million.

Both of the floods outlined above occurred in August, the main month for floods in Britain.

Floods like this are costly in terms of both life and property, and hydraulic engineers have to estimate the recurrence interval of particular flood peaks. Bridges, dams, storm drains, culverts, river channels and even some buildings are planned with specific time periods, related to the recurrence interval, in mind. If the life span of a building is expected to be 20 years, for example, it would be extravagant to build it to withstand the 200 year flood. Large civil engineering structures, such as dams and large bridges, are built with much longer life spans. Failure of such a structure is a considerable potential danger and the safety factor in their construction must be high. Recurrence intervals of 300–400 years must be considered. These time spans are beyond the limits of reliable data, and methods of extending relatively short term information on flood levels have been devised. If the recurrence intervals of a number of discharges for the catchment are calculated as described on page 108, and these are plotted against the discharge on logarithmic graph paper, they form an almost straight line (Fig. 4.21). If the line is extended, it is possible to estimate the recurrence interval of discharges previously unrecorded in the catchment.

Fig. 4.21 Flood recurrence intervals and discharge

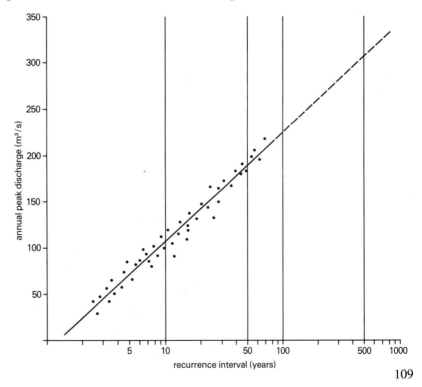

(c) The effect of humans on run-off and flooding

It seems that most human activity has tended to increase run-off. Clearing of woodlands, drainage of marsh areas and ploughing of fields has increased the velocity of surface flow, reducing the time taken by water to reach streams and rivers. Table 4.6 illustrates how the run-off from storms is increased when a catchment is taken over for urban use. Urbanisation increases the area of

Table 4.6 The effect of urbanisation in three catchments in California (Adapted from Waaninen)

Catchment	1961	1962	1963
A	42	31	59
B	46	55	121
C	45	83	206

Run-off in mm rainfall equivalent for three catchments approximately
1.2 km² at Menlo Park, California.

impermeable surface in the catchment. Houses, roads, paths and pavements all feed water into the drainage system, producing rapid run-off in comparison with surrounding vegetated areas. In 1961 none of the three catchments had any urban development. During 1962 and 1963 catchment C became urbanised, followed by B; A remained as a grazing area. Initially all the catchments had a similar total discharge. With urbanisation, the run-off increased by four times in the most heavily urbanised area.

Urbanisation has shortened the recurrence interval of particular discharge peaks by up to half where the density of housing is particularly heavy.

(d) Human response to the effects of flooding

It is possible to reduce the damage produced by floods in two main ways.
(i) *Flood protection* consists of modifications to the channel or banks to enable the river to carry greater discharges. Levees have been built extensively on some rivers, particularly the Mississippi. These artificially raise the bank height so that the depth is increased and greater discharges are accommodated within the river channel. The exceptional flood may still overtop the levee and additional problems are created in returning this flood water to the channel when the river level has subsided, since the river may be higher than the surrounding flood plain.

In Britain the river authorities frequently modify river beds. Large boulders are removed and placed along the banks. This has the double advantage of reducing the bed roughness, therefore increasing the velocity of flow, and protecting the banks from erosion. The maintenance of an efficient channel in this way is very important. Sudden surges during the Lynmouth flood (page 108) were probably due to temporary dams forming from trees and boulders washed down the swollen river. The flood waters built up behind them and suddenly burst through, adding to the effect of the already large discharge.

Dams have been built in many parts of the world to control flooding, though frequently this is only one facet of a multipurpose scheme. The Clewedog dam in Powys has reduced the flood hazard in the lower Severn valley (Plate 4.3) but also supplies water and a recreational area. The dam allows the rate at which water passes down the river to be controlled and thus the flood level can be controlled.

Plate 4.3 Flooding in the Severn Valley near Buildwas, upstream from Ironbridge, Shropshire (Aerofilms)

Diverting rivers away from vulnerable areas and raising the level of the flood plain by dumping material on it are two commonly used flood protection measures. Ash from power stations in the Trent valley has provided landfill for flood plain sites on which factories have subsequently been built.

(ii) *Flood abatement* tackles the problem at source by slowing down the rate at which water from storms reaches the river channel. This has the effect of lengthening the lag period (page 100) and the period of storm flow, and flattening out the storm peak as recorded on the hydrograph. There are several ways in which this may be achieved. Afforestation of the catchment slows down the

111

rate of surface run-off as well as increasing evapotranspiration. Terracing of farm land enables overland flow to be controlled. Contour ploughing and strip cultivation are used in areas of the world where the climate borders on semiarid; they reduce the rate of surface run-off, prevent soil erosion and reduce flood liability.

There is some debate over the relative merits of flood protection or flood abatement. Should the catchment be forested or should a large dam be built in order to control flood water? The debate is often referred to as the 'upstream-downstream controversy', relating to the location of the flood control activity. The problem hinges on whether prevention is better than cure. The answer probably lies mid-way between in some form of co-ordinated catchment management programme.

(iii) In some areas the *erosion of the river banks* has been prevented by reinforcing them with concrete. This is expensive and is largely confined to critical sites, such as urban areas or around bridge foundations where erosion of the bank could prove very expensive. Frequently the banks are protected by removing material from the river bed. The use of large boulders has been mentioned previously (page 110). Smaller sized material is often placed in strong rectangular wire baskets which are then used to face the banks of the channel. Plate 4.4 shows this on the River Dunsop, where the bank has been reinforced to prevent undermining of the road.

Plate 4.4 Bank protection measures in the River Dunsop (Author's photograph)

In areas where the bed load is large and the river prone to flooding, the floor of the channel is sometimes lined with wood. In the Alps, rivers fed by glaciers rise rapidly during the day as the glacier melts. Plate 4.5 shows an Alpine river in flood, the water is densely coloured with sediment and appears opaque. There is a large bed load and the floor of this channel is lined with wooden sleepers. The wood is more elastic than concrete and so resists chipping and abrasion to a greater degree. Wood is also cheaper and more easily repaired than concrete. These factors make its use even more desirable.

Plate 4.5 An Alpine river in flood (A.S. Freem)

(iv) The *deposition of sediment* in the channel is a problem more commonly encountered in the lower reaches of the river or as the aftermath of a flood (page 133). In the lower reaches floods are less efficient in 'flushing' out the channel, as flood discharge is small in relation to the average discharge of the river.

Dredging is frequently needed to deepen the channel for navigation. This is a short term solution, since dredging increases the cross-sectional area and hence lowers the velocity, which increases the tendency of the river to deposit its load.

The *straightening of sinuous rivers* increases the gradient and hence the velocity of flow in the channel. The effect of this may be to increase erosion sufficiently to inhibit deposition in the channel and to maintain the river at a navigational depth unaided (page 130).

(v) *Afforestation and revegetation* of catchments reduce the sediment yield of the rivers. Reservoirs, particularly in semi-arid areas, are very prone to siltation. The Colorado River in the USA flows through desert and semi-desert; the sediment load is high and as a result, dams like the Hoover are rapidly approaching the end of their useful life as sediment fills in the reservoir. In Britain, Thirlmere in the Lake District and Ladybower in the Peak District both have heavily forested catchments to cut down the sediment entering the reservoir.

1. *Discuss how you would apply the ideas of flood abatement and flood protection to the River Dunsop catchment (Fig. 4.8) in the light of the storm outlined on page 108. Bear in mind that it is a rural area with very little settlement within the basin but with a small village at the outlet.*
2. *Table 4.7 gives annual peak discharges for Seneca Creek in the USA. This basin has an area of 260 km². What is the recurrence interval of the highest flood shown in the table? Use the methods described on page 109 to calculate the magnitude of the 75 year flood.*

Table 4.7 Annual peak discharges for Seneca Creek at Dawsonville, Maryland, USA, 1931–1960 (m³/s (Source: Leopold, *Water*, W.H. Freeman and Company. Copyright © 1974)

1931	48.9	1941	36.8	1951	68.5
1932	39.1	1942	41.3	1952	79.6
1933	263.3	1943	102.5	1953	207.5
1934	75.7	1944	75.3	1954	35.1
1935	40.2	1945	59.7	1955	74.2
1936	57.2	1946	83.2	1956	424.7
1937	73.9	1947	56.3	1957	27.2
1938	64.6	1948	56.3	1958	103.1
1939	60.9	1949	63.4	1959	55.8
1940	49.3	1950	64.6	1960	45.3

3. *Describe how you would protect the area shown in Plate 4.3 from further floods like the one illustrated.*

Key Ideas

A. *Introduction: the hydrological cycle*

1. Water is continually entering and leaving the river system which is an *open system* within the larger *closed system* of water circulation, the *hydrological cycle*.
2. Rivers are efficient agents in fashioning the earth's surface.

B. *The drainage basin*
1. Drainage basins form more or less self-contained and convenient units in which to consider the action of running water.
2. Drainage basins can be considered as open systems in which the *inputs* are the various forms of *precipitation*. The major *outputs* are *run-off, groundwater flow, evaporation* and *transpiration*.
3. Since accumulation does not occur in the basin, in the long term inputs are balanced by outputs.
4. The form of river basins can be measured objectively by reference to three axes mutually at right angles, the linear, areal and relief properties.
5. There are fairly constant relationships within an area between *stream order, frequency length* and *basin area*. Some of these have been called 'laws'.

6. Differences in these relationships between areas are responses to changes in precipitation, geology, climate, vegetation and time.

C. *River channels*
1. The *velocity* characteristics of a stream are determined by the total shape of the channel.
2. The rate of conversion of *potential* to *kinetic energy* is determined by the form of the channel.
3. The correlation between the *hydraulic geometry* of stream channels and the *discharge* indicates the close interrelationship of form and process.
4. Variation in discharge results from changes in the precipitation input and the *run-off characteristics* of the basin.

D. *Erosion and transportation*
1. *Erosion* is the removal of material from the bed of the channel.
2. The eroded material is reduced in size during transport by the river.
3. The kinetic energy of a river, measured by its velocity, determines the *competence* of the river to carry its load.
4. The *capacity* of the river to carry its load is mainly a function of its discharge.

E. *Problems in river management*

1. Flooding occurs when the *bankfull stage* is exceeded.
2. The frequency of any particular flood height is denoted by its *recurrence interval*.
3. Most erosion and deposition occurs during flood periods, making flooding one of the most widespread of natural hazards.
4. Frequently, human activity has accentuated the flood hazard.
5. An understanding of the factors which control run-off enables steps to be taken within the catchment to reduce flood heights.
6. Knowledge of the way rivers erode enables the channel to be modified to reduce flood damage.

Additional Activities

1. It was shown on page 99 that width, depth and velocity all increase downstream. Test these ideas by measuring cross sections of the river at a number of points along its length.

   The mean depth can be measured by stretching a line across the river and measuring the depth at regularly spaced points across the channel (Fig. 4.14). The plotting of this data also gives the wetted perimeter and cross-sectional area (mean depth × width).

   Velocity can be measured using floats timed over a measured length. The method gives only surface velocities and it is necessary to apply a correction factor (page 94). Floats which lie low in the water are least subject to disturbance by the wind. Current meters enable velocities at different depths to be measured. In small streams a good indication of the average velocity can be obtained using a conductivity meter. A weak salt solution tipped into the top of a measured section can be detected by a rise in the conductivity. The time of the peak conductivity is a measure of the average velocity. In very small streams V notch weirs can be used (page 95).

2. Rock particles resulting from weathering are often sharply angular, while fluvial erosion tends to round them (page 105) and reduce them in size. Test the effectiveness of rivers in producing rounding by measuring four dimensions of a pebble. The *a* axis is the longest dimension of the pebble. The shortest dimension at right angles to this axis is the *c* axis. The *b* or intermediate axis is the dimension at right angles to the plane of the *a* and *c* axes. The fourth measurement is the radius (*r*) of the sharpest angle in the plane of the *a* and *b* axes. This plane is the one in which the silhouette of the pebble looks largest. These measurements should be made on a sample of at least 50 pebbles chosen within a half metre quadrat from the bed of the channel. The mean length of the *a* or *b* axes can be used to indicate the average size of the sample.

The flatness (*f*) or the shape of the pebble is calculated from the formula:

$$f = \frac{a + b}{2c} \times 100$$

Fig. 4.22 Rainfall and hydrograph data for the storm of 5–6 August 1973 in the River Wye basin (From Newson, M.D. *The Plynlimon Floods of 5–6 August 1973*, Institute of Hydrology, Wallingford, 1975)

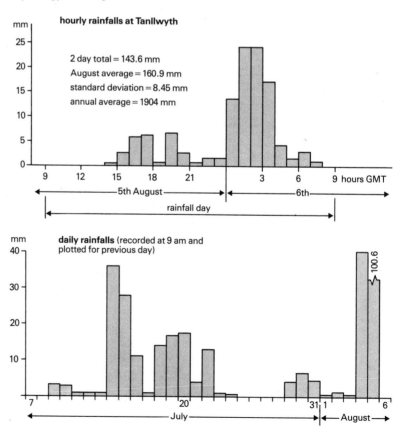

116

Values of 100 indicate pebbles in which a, b and c are of equal length. Higher values indicate increasing flatness.

The roundness index (R) measures the sharpness of the edges of the pebble sample. It is calculated from:

$$R = \frac{2r}{a} \times 1000$$

Values of 1000 indicate perfectly rounded pebbles, while lower values are produced by more angular pebbles. Both flatness and roundness should be sample means.

Changes in the values obtained for mean length, flatness and roundness occur both downstream and across the channel. These variations reflect changes in the velocity of flow in the channel. There are also strong influences on pebble shape exerted by rock type. Some lithologies are very resistant to abrasion and if each major lithology is studied separately, rates of rounding along the channel can be compared. It is possible to remove the effect of lithology from your study by measuring only one of the lithologies in the channel.

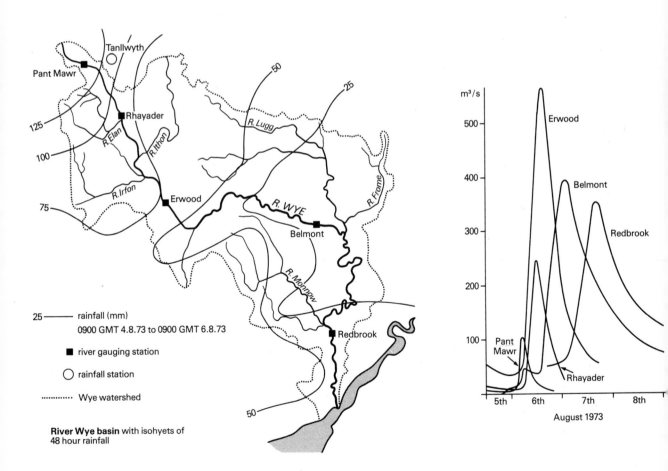

River Wye basin with isohyets of 48 hour rainfall

117

3. Discuss the factors which cause the discharge of rivers to vary over time.
4. Assess the relative importance of weathering, slopes and rivers in removing rock material from the land surface.
5. Study carefully the map, hydrographs and rainfall data for the River Wye basin given in Fig. 4.22.
   (a) Explain how the rainfall during the period 7 July to 4 August would have influenced the run-off from the storm of 5 and 6 August.
   (b) Compare the hydrographs for Pant Mawr and Rhayader which resulted from the storm of 5 and 6 August. Explain the differences you have observed.
   (c) Explain the differences in the hydrographs for Erwood, Belmont and Redbrook.

Reading

*BRUCE, J.P. and CLARKE, R.H., *Introduction to hydrometeorology*, Pergamon Press, Chapters 4 and 8, 1966

*CHORLEY, R.J., *Introduction to fluvial processes*, Methuen, pages 74–156, 1971

*COOKE, R.U. and DOORNKAMP, J.C., *Geomorphology in environmental management*, Oxford University Press, Chapters 1 and 5, 1974

*LEOPOLD, L.B., *Water*, W.H. Freeman, Chapters 3–5, 1974

LEOPOLD, L.B. *et al.*, *Fluvial processes in geomorphology*, W.H. Freeman, Chapters 5 and 6, 1964

NEWSON, M.D., *Flooding and flood hazard in the United Kingdom*, Oxford University Press, Chapters 1, 4 and 5, 1975

SMITH, D.I. and STOPPS, P., *The river basin*, Cambridge University Press, 1978

STRAHLER, A.N., *Physical geography*, Wiley, pages 413–437, 455–470, 1975

WEYMAN, D.R., *Run-off process and streamflow modelling*, Oxford University Press, Chapters 1–4, 1975

WEYMAN, D.R. and WILSON, C., *Hydrology for schools*, Geographical Association, 1975

5 Fluvial landforms

In the previous chapter we studied the patterns which many rivers and streams seem to follow and looked at the various processes involved. In this chapter we go on to examine the effects such fluvial processes have on the physical landscape.

A. Long Profiles

The long profile of the river covers the changes in altitude from the source along its channel to its mouth. In the previous chapters we saw that the headwaters of a river, first order tributaries, have steeper gradients than those of higher orders (page 88). The long profile of the River Ribble (Fig. 5.1) shows steeper headwaters and progressively lower gradients as the river approaches sea level. This section is concerned with how that generally *concave* profile is determined.

Fig. 5.1 River Ribble long profile plotted logarithmically and arithmetically

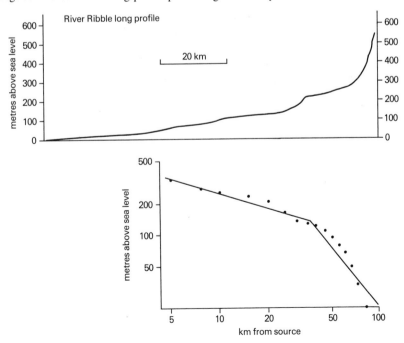

Many long profiles are not smoothly concave but consist of a number of concave sections. It is possible to examine the regularity of these sections by plotting them to logarithmic scale. Plotting the height scale logarithmically renders smoothly concave sections as straight lines. The River Ribble when plotted like this shows two straight sections. Such profiles are termed polycyclic. The term originated at the time when long profiles of rivers were all thought to show the effects of multiple changes of sea level at rejuvenations (page 136). In Chapter 7 the detailed history of the Ribble valley is examined. This explains the complexity of the long profile in another way (page 189).

1. Downstream changes in the channel

On page 101 the discharge of rivers was shown to increase with increasing distance from their source. The extra water comes from tributaries which increase the drainage area of the river. Figure 4.16 showed that the depth, width and velocity also increase downstream. Any increase in mean depth and width must also increase the hydraulic radius. This implies that the river flows more efficiently as the channel becomes larger, since there is relatively less contact of water with the bed (page 96).

The effect on the long profile of this increasing efficiency of the channel downstream can be shown by referring to Manning's equation, page 97. If we take the part of this equation that represents the velocity:

$$V = \frac{R^{\frac{2}{3}} \times S^{\frac{1}{2}}}{n}$$

where V represents velocity, R the hydraulic radius and n the bed roughness, then assuming that n remains constant, any increase in R downstream will tend to increase the velocity, indicating that the channel is more efficient. We already know from Fig. 4.16 that in most natural channels the mean annual current velocity does increase downstream. In reality the increase in the hydraulic radius, and so velocity, together with the larger cross-sectional area which must occur when depth and width are greater, more than compensate for the increase in discharge. The channel adjusts to this by a decrease in the angle of the slope in the downstream direction. Upstream sections of the river therefore have steeper channel gradients than downstream sections, and the profile is concave.

The concavity cannot be due to increases in the discharge alone, since rivers like the Nile and the Colorado, whose lower courses are in deserts, still have concave long profiles. Nor does the age of a river channel seem to have a significant influence, since rills developed on freshly exposed surfaces also show concave profiles.

In most rivers the size of sediment which lies on the bed decreases downstream as a result of the attrition (page 105). This will tend to make the bed smoother and so a decrease in bed roughness, Manning's n value, could be expected, which is a further factor in increasing the velocity downstream.

The arguments above refer to the 'average' river, the mean of many measurements made on a large number of rivers. It is always possible to find exceptions where the profile is convex, as are parts of the profile of the River Ribble (Fig. 5.1). If we wish to explain these in the manner outlined above, all we need seek are reasons for a *decrease* in hydraulic radius downstream. For example, in an area of hard bedrock vertical erosion may predominate, producing a steep-sided, narrow gorge and a lower value of hydraulic radius. The efficiency of the channel is diminished. Convexity of the long profile can therefore be expected in an area where the river crosses hard bedrock. Convexities usually mark the site of rapids and waterfalls (page 122).

2. Thermodynamic principles

Another way of looking at the problem of the concave nature of the long profile is to use two of the principles of thermodynamics: the principle of least work, and the principle of uniform distribution of work.

The work done is the energy expended in travelling from the source of the stream to sea level. The principle of least work would give a waterfall from source to sea level. All the energy would change instantly from potential to kinetic (page 92). The principle of uniform distribution of work would give a straight line profile from source to mouth. But since discharge in rivers increases downstream (page 101), a gently concave profile could be expected. The actual outcome of these conflicting principles in reality gives a profile mid-way between the two extremes (Fig. 5.2). In reality you will see from the profiles in the assignments at the end of this section (page 124), that few rivers have perfectly smooth concave long profiles.

Fig. 5.2 The principles of thermodynamics and the long profile

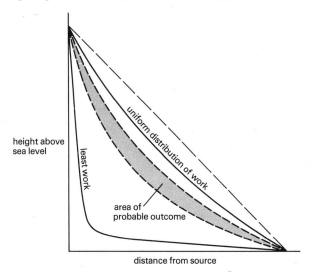

3. The concept of grade

On page 67 we considered slopes in which inputs and outputs from a segment were in equilibrium or in a steady state. The same ideas can be applied to a reach of a river. A *graded reach* is one in which the slope of the channel is adjusted so that there is no net gain or loss of sediment from the reach. All the kinetic energy of the river is used to transport sediment. There is no excess available for erosion. Nor is there an energy deficit which would cause the sediment load to be deposited.

If a river is not able to carry all the load entering a particular reach, some must be deposited. This build-up of sediment steepens the bed gradient and so increases the velocity to a point where all the load can be transported and the equilibrium is restored. Should the river be more than able to transport the sediment entering the reach, then sediment is eroded and transported. This erosion reduces the gradient, and thus velocity, until the reach returns to the equilibrium state or grade.

The idea of self-regulation of the long profile by the river is dependent on a *negative feedback* mechanism (page 10). This is simply a change in one part of the system which causes a change in the opposite direction in another part of the system. In this case a decrease in stream velocity causes an increase in stream gradient.

The state of equilibrium or grade constantly changes with fluctuations in discharge. The channel may be continually adjusted to new sets of conditions which tend to vary about the 'average' condition of the river. In such a situation the channel is said to be in *dynamic equilibrium*.

4. Irregularities in the long profile

(a) Waterfalls and rapids

Rivers frequently flow across geological boundaries and at these points the river picks out the differences in the resistance to erosion of the rock types.

Fig. 5.3 Thornton Force, waterfall formation and retreat

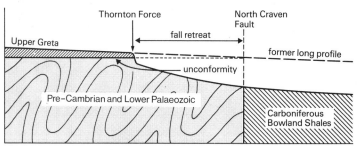

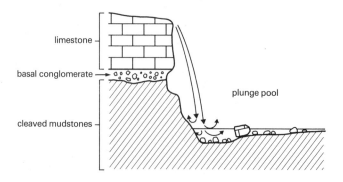

Where the river flows from hard to soft rock, the softer is eroded relatively rapidly and the gradient is locally steepened to form a waterfall or section of rapids. Figure 5.3 shows the effect of this at Thornton Force, near Ingleton in West Yorkshire. The upper part of the fall is in the lower levels of the Carboniferous Limestone. At the base of the limestone is a thin, irregular stratum of conglomerate containing pebbles up to 200 mm in length which rests on much older folded and cleaved mudstones and shales. The shales have been rapidly eroded because of the presence of the closely spaced cleavage planes. The upper surface of the shales was weakened by exposure to weathering before the gravel of the conglomerate was deposited. The weakened rock has been picked out to form the shallow cave at the back of the fall. The water now falls about 10 m and has excavated a deeper plunge pool at the foot of the fall. This waterfall has retreated from the line of the North Craven Fault (Fig. 5.3), leaving a deep valley which exposes the shales underlying the limestone in this area.

In most rivers changes of slope at waterfalls and rapids cause an uneven distribution of energy along the long profile and hence erosion of the feature and eventually a lowering of the gradient.

Plate 5.1 Thornton Force (Author's photograph)

(b) Pools and riffles

Even where the bed is composed of alluvial material, inequalities still exist and are marked by shallower, faster flowing areas, *riffles*, and deeper, slower flowing pools (Fig. 5.4). The riffles mark areas where energy is being used rapidly

Fig. 5.4 Energy accumulation and loss in a channel of varying slope

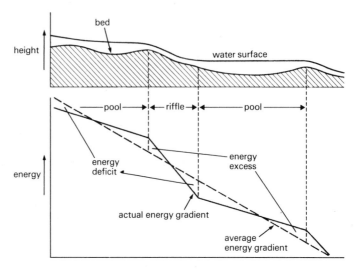

to erode the material of the river bed. The pools are areas where the river builds up energy above the average energy gradient of the river. This energy is dissipated by the next riffle in the river. On a small scale the pattern of energy dissipation along the river is far from uniform. The riffles and pools affect the water surface most noticeably when the water level is low. At higher levels the gradient of the water surface is more uniform, as is energy dissipation. The total energy is also much greater.

Table 5.1 Long profile of the River Teifi, Wales

Height (m O.D.)	Distance (km)	Height	Distance	Height	Distance
15	10.0	198	2.3	381	0.20
30	11.6	213	0.6	396	0.10
46	6.0	229	0.3	411	0.60
61	5.5	244	0.25	427	0.30
76	6.5	259	0.20	442	0.20
91	7.5	274	0.35	457	0.20
107	11.0	290	0.25	472	1.50
122	10.5	305	0.20	488	0.25
137	4.3	320	0.15	503	0.10
152	9.3	335	0.25	518	0.25
168	14.5	351	0.10	533	0.25
183	1.5	366	0.15	549	0.20

124

1. (a) *Using the data in Table 5.1 on the long profile of the River Teifi, plot height against distance, using: (i) arithmetic scales; (ii) a logarithmic scale for the height (page 119); (iii) a logarithmic scale for both height and distance.*
   (b) *What are the advantages and disadvantages of each of these methods representing the data?*
   (c) *From the profiles what can you deduce about the development of the River Teifi?*
2. *On a tracing paper overlay of Plate 5.1, trace the outline of the limestone, conglomerate and shale beds. Label these, together with the plunge pool cave and any limestone blocks which have fallen from the lip of the fall (Fig. 5.3).*
3. *The three boxes in Fig. 5.5 represent erosion, channel slope and stream velocity. The relationship between one pair of these variables is indicated by the line in graph A. What are these variables? Complete the other box and the graphs and identify the negative feedback in this system.*

Fig. 5.5 Systems diagram relating hydraulics variables

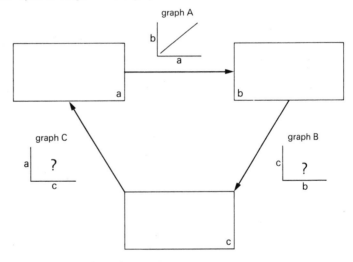

B. Transverse Profiles

1. Vertical and horizontal erosion

We have seen that the principal work of stream and river erosion is to deepen the valley vertically. However, if vertical erosion were the only process operating, every stream would be flowing at the bottom of a narrow gorge with vertical sides whose width was the same as the stream. Although this does occur under certain conditions, for example, where the stream has cut a valley in a very hard rock, weathering and other slope processes (see Chapters 2 and 3) reduce the angle of the valley sides to less than vertical.

However, the angle of the valley sides varies greatly from area to area and with increasing distance from the source of the stream. The cross-profile of a stream valley thus varies from narrow, steep-sided gorges to gentle, open form. Valleys may also be asymmetrical, slope angles differing from side to side. Nor is the slope angle necessarily constant over the whole length of the slope.

2. Connections with weathering and slope systems

It is once again convenient to think of this as a 'system', with an input and output (Fig. 2.4).

From Fig. 5.6 it is clear that a certain amount of the weathered debris supplied to the bottom of the valley sides by slope processes is removed by the stream, the actual amount depending on the stream velocity and discharge.

Fig. 5.6 The slope system river erosion and transport

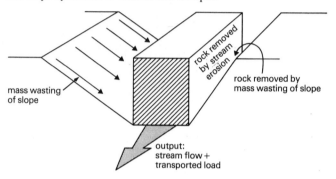

Thus we can refer to the efficiency of the stream action in this context in relation to how much or how little weathered rock debris is removed. Whilst the finest particles may be removed under normal conditions, most of the debris will be removed only at times of peak discharge under storm conditions. Dry valleys in limestone districts, such as above Malham Cove in Yorkshire, show a very inefficient system since there is no surface stream flow at all and the weathered limestone collects on the valley floor, the only removal being by rainwater solution.

Generally, however, valley slopes will be steeper if the stream is actively eroding, if the solid rocks are hard and resist erosion, or if the stream is involved in lateral erosion and is trying to undercut the base of the slope (Plate 5.2).

ASSIGNMENTS

Using Fig. 5.6, explain how the relative rate of stream erosion affects the form of valley slopes.

C. Rivers in Plan

1. Sinuosity

In the discussion on long profiles of rivers no mention was made of the directness of the river channel. Few river channels are exactly straight. The degree to

Plate 5.2 Steep valley slopes in Settle Beck Gill (Author's photograph)

Fig. 5.7 Sinuous and meandering river channels

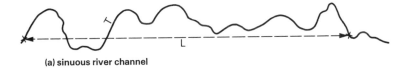

(a) sinuous river channel

(b) meandering river channel

which they vary from a straight line is termed sinuosity and is measured by the ratio of the channel length to the straight line between the same two points (Fig. 5.7):

$$\text{Sinuosity} = \frac{\text{channel length}}{\text{straight line distance}} = \frac{T}{L}$$

River courses are frequently regularly sinuous or *meandering*. In connection with river energy the increased length in a sinuous channel means that the same amount of energy is spread along the greater length. The river may be approaching a 'graded state' and vertical erosion is limited. The available energy is used in lateral erosion producing meanders. There appears to be a limit to the lengthening of the river by these methods.

2. Geometric properties of meanders

The geometric variables of meanders, such as width and wavelength, shown in Fig. 5.8, have been investigated in relation to other factors such as discharge and sediment size. Some of these relationships are illustrated below. They have increased our understanding but not completely solved the enigma of the almost mathematical regularity of some meanders.

Fig. 5.8 Geometric measurement of meanders

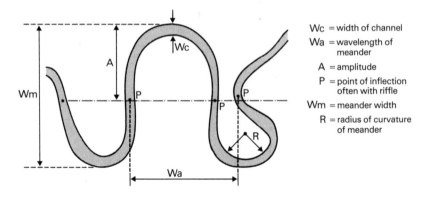

Wc = width of channel
Wa = wavelength of meander
A = amplitude
P = point of inflection often with riffle
Wm = meander width
R = radius of curvature of meander

The constancy of the ratio between some of these geometric values reflects the regularity of meanders:

$$
\begin{aligned}
\text{Wc/R} &= 2 - 3 \\
\text{sinuosity} &= 1.5 - 4 \\
\text{Wc/Wa} &= 7 - 10 \\
\text{T/Wc} &= 11 - 16
\end{aligned}
$$

The wavelength of meanders shows a close correspondence (correlation) with mean annual discharge, more so than with bankfull discharges. It therefore appears likely that the meander form is determined by the average flow of the river. Observation of the flow in meanders during flood reveals a straighter path of the maximum velocity channel than in low flow conditions.

Some meanders show a wavelength far too large for the discharge they carry at present. This has led to the idea that they are related to periods of greater dis-

128

charge in the past, such as the melting of the Pleistocene ice sheets, rather than to present conditions. They can be considered as 'fossil' features of the landscape, and can be identified as *anomalies* on a graph plotting meander wavelength against discharge (Fig. 5.9).

Fig. 5.9 Meander wavelength and discharge (After Leopold *et al.*)

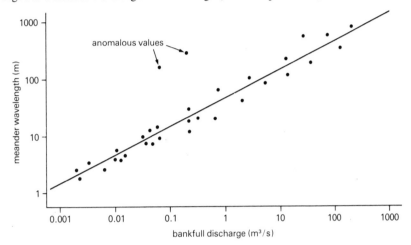

3. Flow in Meanders

Understanding the origin of meanders is problematical. In low flow conditions straight channels in alluvial material are seen to have alternating bars of sediment in the river bed so that the flow is forced to weave around them (Fig. 5.10d). These are the *riffles* described on page 124. Between them lie deeper areas, the *pools*. The material which makes up the riffles may vary in size from silt to coarse gravel, depending on the dominant velocity in the channel and the nature of the rock in the channel.

The swing of the flow induced by the riffles tends to direct the zone of maximum velocity and deepest water to one side of the channel, producing erosion (Fig. 5.10b and c). This concave bank is eroded by undercutting of the alluvial material of the flood plain. Where the bank is of silt or clay a steep cliff results since the particles are cohesive. Sand and gravels form lower angled slopes since they tend to roll or flow instead of collapsing as a block into the channel.

Erosion of the concave bank is accompanied by deposition on the convex bank, forming a sandy or gravelly area called a *point bar*. The width of the channel remains relatively constant, while the whole channel migrates laterally and becomes more sinuous, acquiring the typical meander shape described on page 128.

Water flow in meanders appears to be *helicoidal* as illustrated in Fig. 5.10a. The water flows fastest on the outside of the bend in an attempt to maintain a straight course. It is deflected round the bend by the river bank. The effect of this is to produce a surface flow towards the outside of the bend and a compensatory return flow across the bed while the water moves down channel. This is

Fig. 5.10 Flow in meanders: a) helicoidal flow in a curved channel section, b) flow transverse to the main flow in a meander illustrating the lateral component of the helicoidal flow, c) position of main flow in meander and d) low water flow in a straight segment

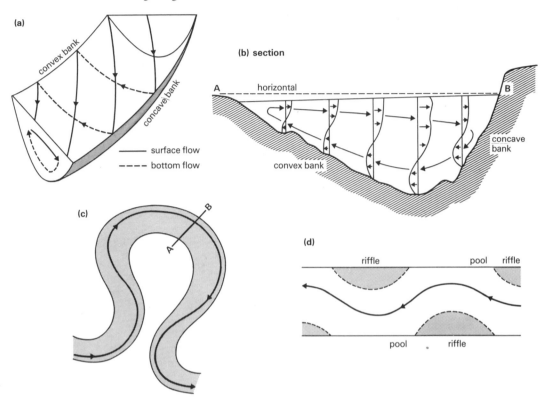

Fig. 5.11 Channel movements of the Mississippi: the 1881 and 1930 surveys show the meanders downstream; the meander around Moss Island was cut off in 1821 (After US Army Corps of Engineers, in Strahler, *Physical Geography*, Wiley)

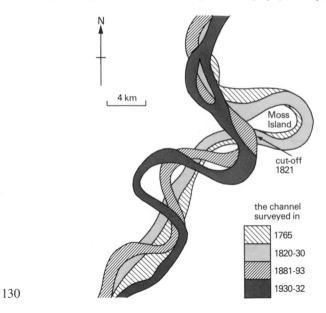

seen in Fig. 5.10a, where the flow is resolved into two components: (i) down channel, and (ii) at right angles to (i) across the channel.

It is improbable that any single particle of water completes this helicoidal course. This helicoidal flow is found in nearly straight channels and the question of cause and effect still remains to be answered satisfactorily.

The effect of helicoidal flow is to transport sediment eroded from the concave bank downstream to the convex bank where it is deposited in the slack water to add to the point bar.

The greatest erosion of the concave bank occurs just downstream of the axis of the meander bend, because the course of the maximum velocity zone in the channel does not perfectly reflect the meander shape (Fig. 5.10c). This causes the meander to migrate down the valley. Rates of downvalley movement of 15 m/a have been recorded in Russia. The section of the Mississippi shown in Fig. 5.11 shows meander movements downstream. The cut-off which occurred

Plate 5.3 Meandering river with cut-offs, oxbows and scrolls. The River Clyde near Carstairs, Scotland (John Dewar Studios)

in 1821 considerably accelerated this rate and the later surveys show the progressive migration of the meander form down the valley.

If sinuosity becomes too great, the river in flood may be able to break through the meander neck (Fig. 5.11), thereby straightening its channel, increasing its gradient and reducing energy loss. The abandoned channel may be used during flood but frequently the ends nearest the new course become blocked with alluvium and the channel becomes a small lake or *oxbow*. These are rapidly filled until only marshy tracts are left. Such indications of the former course of the river are termed *meander scrolls*. They are shown on Plate 5.3.

Meanders are not confined to alluvial channels, they also occur in ice on glaciers where water flows down the surface, in caves where solution is a major erosion process and in the Gulf Stream.

ASSIGNMENTS

1. *Figure 5.12 shows part of the River Dee where it forms the border between Wales and England.*
   (a) *What evidence is there that the course of the river has changed? Trace the present and the past course of the river.*
   (b) *Calculate the sinuosity of each of these two courses from A to B (page 128).*
   (c) *What conclusions can you draw from these measurements about past and present discharges?*
2. *Refer to Plate 5.3 which shows a meandering river and tributary. Place a sheet of tracing paper over the plate. Identify, draw and label as many physical features as possible which contrast the tributary river with the main river.*

Fig. 5.12 The River Dee near Holt

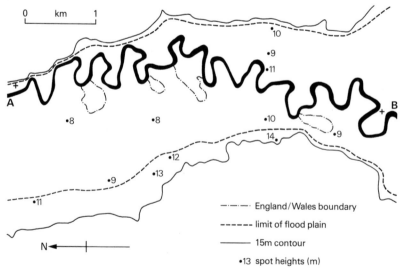

3. *Refer to Fig. 5.11. Calculate the rate of downstream meander migration: (a) between 1765 and 1820, and (b) between 1881 and 1930. The late nineteenth and early twentieth century was a period of rapid colonisation of the Mississippi basin. Bearing this in mind, can you explain the differences between (a) and (b)?*

D. Flood Plain Formation

It has been shown how meanders migrate both downstream and laterally as a result of erosion of the concave bank. Although they are more often found in valleys floored with alluvium where they rework previously deposited sediment, they can form in consolidated rocks. The processes of erosion are much slower, and the movement downstream of the channel shape widens the valley floor and deposits the material forming the flood plain. The River Dee on the border between England and Wales (Fig. 5.12) shows how the flood plain may be wider than the *meander belt*. The narrow zone occupied by the meanders is confined to the east side of the flood plain.

1. Flood plain deposits

In detail the alluvium which forms the flood plain originates from two sources: (i) from overbank deposition which occurs at time of flood, and (ii) from channel deposits, mainly point bar deposition in migrating meander forms. The former are generally fine sands, silts or clays, while the point bar sediments are gravels or coarser deposits. In a large number of rivers the frequency with which the river overflows its banks is at a recurrence interval of between one and two years. Overbank deposition is of infrequent occurrence in most rivers and addition of material to the flood plain is probably of the order of a few centimetres in 100 years. In tropical areas where the clay content of flood water is high and the flood plain heavily vegetated, accretion rates of between 100 and 200 cm per 100 years are recorded; the Nile adds about 10 cm and the Little Missouri 100 cm in 100 years.

When the river overtops its banks the channel suddenly expands to the full width of the flood plain (Fig. 5.13) and the velocity of the water outside the channel falls considerably. Even the very finest sediment may be deposited as the water becomes static in isolated pools (such as old cut-offs) in the flood plain.

Fig. 5.13 The formation of the flood plain

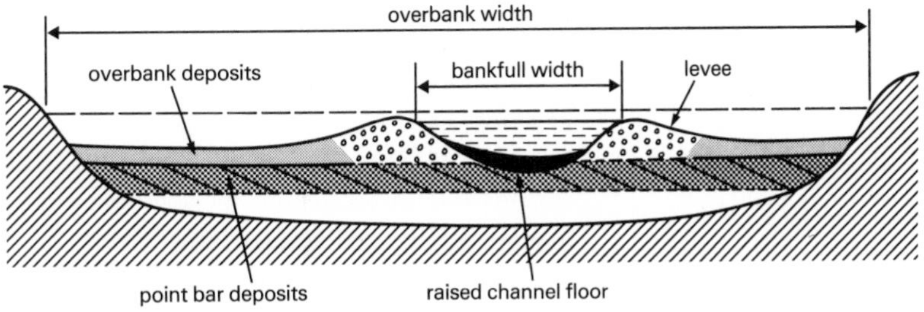

2. Levees

The flood plain immediately adjacent to the channel often consists of coarser sediment and is raised in comparison to the rest of the flood plain. These *natural levees* mark the zone where there is the greatest change in velocity during floods. The coarsest and greatest volume of the load is deposited to form an embankment which effectively raises the height of the bank. The floor of the channel becomes built up between the levees and may be higher than the surrounding flood plain. The levees also prevent water returning to the channel flood. This adds more sediment to the flood plain as the water stagnates and percolates into the flood plain or evaporates.

Plate 5.4 Artificially constructed levees on the River Clwyd, North Wales (Author's photograph)

The height of the levees varies from a few centimetres to several metres on the world's largest rivers. As a flood protection measure the height of levees is often raised, but provision must be made to allow the flood water which overtops the levee to drain back to the channel (Plate 4.3).

3. Braided channels

Some rivers may be supplied with large loads of sand and gravel by their headwater streams. Since sands and gravels are coarse they constitute the bed load. As a result the channel often becomes wide and shallow as the river spreads out to create the most efficient way of transporting this material by making a large bed surface area.

If the gradient of the river lessens and the energy available for transport is reduced (page 121), the river becomes overloaded and deposition occurs within

134

the channel itself. The river divides into several small channels around the small 'islands' of deposits, coming together and dividing again from place to place. In plan view the channels resemble the braided strands of a rope and consequently the name *braided channel* is often used to describe this feature (Plate 5.5).

At times of low discharge a braided river may disappear entirely as the water sinks into the loose sands, gravels or pebbles and flows below the surface. In contrast, in peak discharge the whole braided section may be covered with water and much movement of deposits occurs, often greatly altering the pattern of stream channels when lower discharge conditions once again resume.

Plate 5.5 A braided channel in a river in Arctic Canada (Geological Survey of Canada, Ottawa)

4. Changing conditions in the flood plain

(a) Terraces

Flood plains have so far been related to the conditions responsible for their formation. But some were formed thousands of years ago by conditions which no longer prevail. They are fossil features, like the meanders which wind across them (page 129). Fluvio-glacial meltwaters of the last glaciation increased the discharge and sediment load of many rivers in temperate latitudes and built up the flood plain to heights seldom reached by today's flood conditions. Often the river can be seen to be cutting down through these deposits and the migration of the meanders, forming a new flood plain some metres below the old one. Remnants of these former flood plains are called *terraces* (Fig. 5.14). The Thames has terraces which can be dated, from the fossils they con-

Fig. 5.14 River terraces

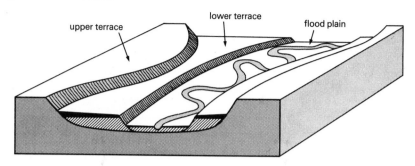

tain, as almost two million years old (page 204). Since that time the river has changed its course but its terraces enable its profile to be reconstructed. Figure 5.15 shows the terraces which lie on either side of the present course of the river. The oldest and highest, the Boyne Hill Terrace, indicates warm conditions of an interglacial period, while the Taplow and the Buried Channel formed during glacial periods. The Buried Channel is cut below present sea level and has since been filled. At the time of its formation, sea level was lower and the river was actively lowering its bed towards this level (page 270).

Fig. 5.15 Cross-section of the Lower Thames terraces

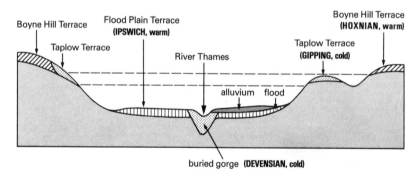

(b) Rejuvenation

The river can cut its bed no more than a few metres below sea level. It is referred to as *base level*. Changes in this base level are common and falls of sea level induce erosion in the lower reaches of the river.

A fall of sea level increases the potential energy of the river at its former mouth (Fig. 5.16). This extra energy enables it to resume more active erosion. The river is said to be *rejuvenated*. The lowest reaches of the river are the first to become adjusted to the new conditions. The adjusted portion extends slowly headwards by eroding and lowering waterfalls and rapids (page 122). The headward limit of this rejuvenation is marked by a change in slope, the *knick point* (Fig. 5.16).

Ultimately the knick point is eliminated and the river is *graded* throughout its length to the new sea level. The rate at which this occurs is very slow, so that

136

Fig. 5.16 Sea level changes and the effect on the profile

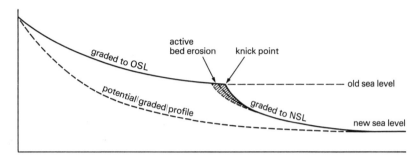

the river may be responding to a fall in sea level which occurred at its mouth half a million years ago and in which time there have been several changes in sea level and climate. The effect of climatic changes on rivers is examined in Chapter 7. Tectonic movements may cause rejuvenation by raising the land. The effects are similar to those due to fall in sea level, though they are on a more local scale.

The changes in gradient in the long profile of a river such as those shown in Fig. 5.16 are not always the result of rejuvenation. Hard rock bands which cross the valley are lowered only slowly and the channel steepens as it crosses them (page 121). This results in two concave parts to the long profile, the upper one graded to the top of the rock barrier. Lakes also act in this way, providing *temporary base levels* below which the river cannot lower its bed.

(c) Incised meanders

Where a river was meandering and its banks were highly resistant to erosion, downcutting would predominate and the meanders would become *incised*. Their shapes remain approximately constant while this occurs. If downcutting and lateral extension occur together, the meanders increase in sinuosity whilst being incised. This type of feature is termed an *ingrown meander* (Fig. 5.17). It is recognised by gentle slopes in the meander cores compared to the steeper slopes on the outside of the meander loop.

Fig. 5.17 Incised and ingrown meanders

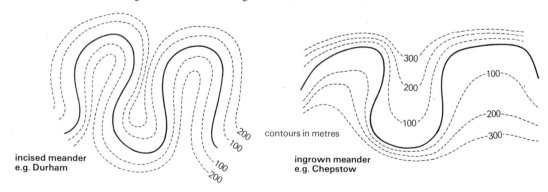

137

(d) Deltas

In the lowest reaches, most rivers are fully laden and the transportation of alluvial material, especially sand, silt and clay, takes up all the energy available. However, when the river reaches the sea its velocity, and therefore competence to transport, are reduced. Deposition occurs in the estuary and out to sea, as far as the river's current may flow (page 274). Where tidal currents flow strongly into and out of the estuary, they have a scouring action and may prevent the alluvial material from being deposited until considerably further out to sea.

Under certain conditions and especially where rivers discharge into tideless or almost tideless seas, the build up of alluvium in the wide channel may cause the river to divide into a series of smaller channels called *distributaries*. Often the shape caused by this deposition resembles the Greek letter 'delta' ($\triangle$) and the feature is called a *delta*. There are two commonly recognised morphological types: *arcuate*, as at the mouth of the Nile; and *bird's foot*, as at the mouth of the Mississippi (Fig. 5.18), where the river has continued to build its levees out

Fig. 5.18 Changes in the distributaries of the Mississippi Delta: channel A was probably used 3000 years ago, B 1500 years ago, C 1000 years ago and D 700 years ago; the modern river channel is about 400 years old

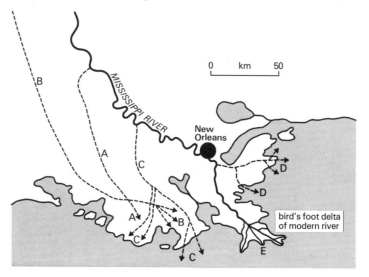

into the sea. The formation of the latter is thought to be the result of river water being less dense than sea water. The river water flows over the salt water surface, slowly spreading out. Alluvial material is deposited at the edges of this flow. Arcuate deltas seem to form when the river and sea waters are of similar densities.

In the analysis of deltaic sediments it is found that the coarsest particles settle first and these horizontal layers are called *top-set beds*. Moving seawards are the *fore-set beds*, which grade into nearly horizontal *bottom-set beds* on the sea floor (Fig. 5.19). When the angle of rest for the fore-set sediments is exceeded by the addition of further sediments, they slide down over the fore-set beds as a

Plate 5.6 The delta of a small river entering Lake Thun, Switzerland (Aerofilms)

Fig. 5.19 Structure of a simple delta (After Gilbert)

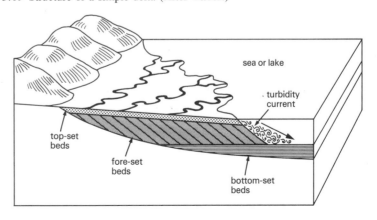

turbidity current. Such currents are able to transport sediments for long distances, sometimes even beyond the limits of the continental shelf.

The process of delta formation is, of course, not limited to the sea. It may occur when a river flows into a lake lying in its long profile, thus checking the velocity of flow and resulting in deposition.

E. Evolution of Drainage Patterns

The arrangement of the streams and tributaries of a river drainage basin is called the drainage pattern.

1. Descriptive patterns

Names which have been given to several recognisable types include *dendritic, trellis* and *radial*.

(a) Dendritic

Perhaps the most common is the dendritic pattern, the network of streams being likened to the branching of a tree (Fig. 5.20). This occurs where geological structure exerts little influence on the course of the river. The River Dunsop (Fig. 4.8) is an example of this pattern.

Fig. 5.20 Drainage patterns

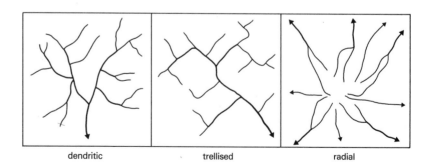

dendritic   trellised   radial

(b) Trellis

The trellis pattern is used to describe a river network whose tributaries join each other at approximately right angles under the control of the geological structure (Fig. 5.20). Rivers which flow in the same direction as the geological dip of the rocks are called *dip* (or *consequent*) rivers. It is assumed that uplift and tilting of the rocks were responsible for determining the principal direction of drainage. A tributary joining at approximately right angles is called a *strike* (or *subsequent*) river. Geologists define strike as a line at right angles to the direction of dip of the rocks. The Medway basin illustrates how such patterns evolve (Fig. 5.21). The consequent streams (1), developed on the original domed upper surface of the chalk, became the major drainage lines. On this uniform surface the drainage may have been almost dendritic with elongation in the direction of the dip of the rocks. Erosion by these rivers removed the chalk cover and the streams began to exploit the lines of least resistant strata. Strike

Fig. 5.21 Geological structure of the Weald and drainage pattern

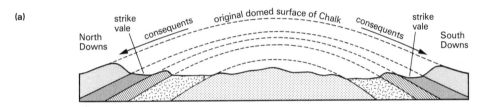

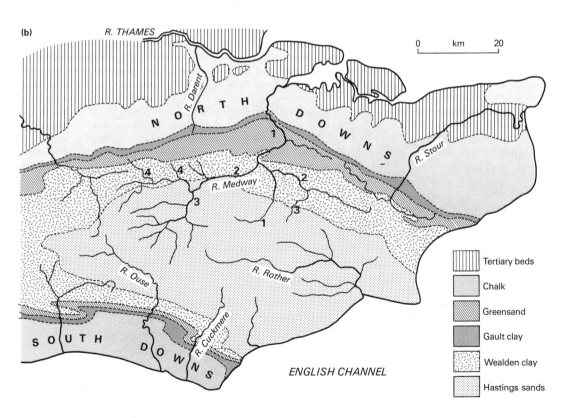

141

streams (2) developed rapidly along the soft Weald Clay Belt. These are fed by a second generation of dip streams (3), developed on the Hastings Beds sandstone layers, while other shorter streams flow into the strike valleys in directions opposite to the dip (4).

(c) Radial

Radial drainage occurs on an area whose rocks have been uplifted into a dome structure and the resulting streams and rivers radiate out from the central high point like the spokes of a wheel (Fig. 5.22). Volcanic cones show this pattern very clearly.

2. Inherited drainage patterns

In addition to the above descriptive patterns, two inherited patterns can be recognised: *superimposed* and *antecedent* drainage.

(i) *Superimposed drainage* is recognised when a previously established drainage pattern is let down or laid onto structures which were originally concealed by later rocks. The English Lake District is a classic example where a radial pattern was established on a former uplifted dome, much of which is now eroded away (Fig. 5.22). The pattern is now superimposed onto the present day

Fig. 5.22 Superimposed radial drainage in the Lake District

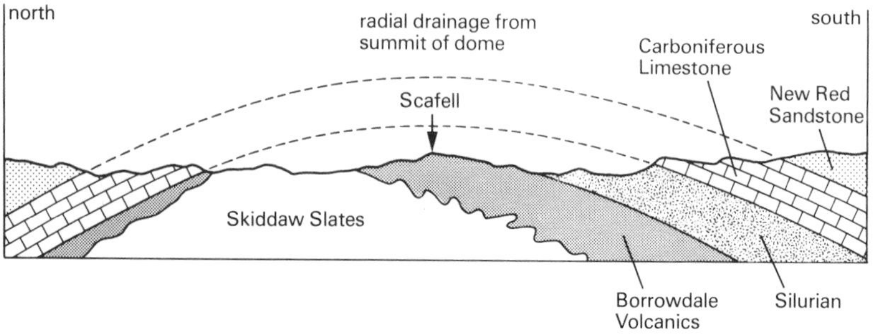

surface whose exposed rocks bear little relationship to the drainage pattern. Glaciation has further emphasised this superimposition by erosion of the valleys during the Pleistocene period. The Ordnance Survey Tourist Map of the Lake District shows the radial pattern of the several ribbon lakes lying within the glaciated valleys.

(ii) *Antecedent drainage* occurs when an established river, existing before orogenic uplift, keeps pace by erosion with the elevating upland and thus cuts a gorge through the mountain chains. The drainage is so called because it is older than the more recent uplift. Antecedent drainage is exemplified in the Himalayas where the Indus and the Brahmaputra Rivers rise in the less mountainous areas to the north and both flow south through gorges in the main Himalayan chains. As an example of the depth of these gorges, the floor of the Indus in Kashmir (Fig. 5.23) is only 1000 metres above sea level. Thus the

Fig. 5.23 Antecedent drainage of the Indus, Brahmaputra and many headwaters of the Ganges

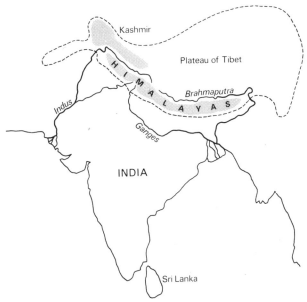

river eroding vertically has kept pace with about 5000 metres of uplift during the Tertiary orogeny when the formation of the Himalayas began.

3. Basin area changes: river capture

Mention has already been made of the slow headward extension of a river valley (page 78). This is caused by a process known as spring-sapping (Fig. 5.24) where at the point of the spring the outflowing groundwater erodes material around it and thus the river cuts back slowly into the hillslope behind

Fig. 5.24 Headward extension of the stream channel by spring-sapping

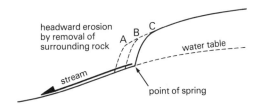

the spring, X in Fig. 5.25. Chance factors may cause a river to retreat headwards in such a way that it 'captures' part of another higher river, and causes that river, upstream of the point of capture, to alter its course and flow down into the drainage basin of the 'capturing' river. At the point of capture a distinct change of direction in the higher captured river will occur, called the *elbow of capture*, Y in Fig. 5.25. Diversion of the additional water from the captured river into the lower river valley creates increases in discharge and thus

Fig. 5.25 River capture

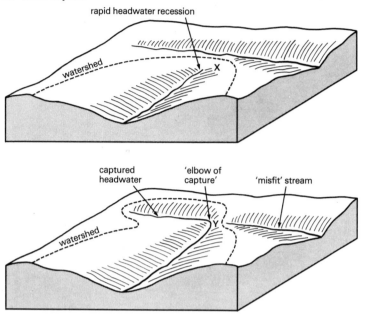

increases the river energy enabling it to rejuvenate its valley by downward incision of the river bed.

Figure 5.26 is an example of river capture where the three main rivers of the North Tyne system clearly correspond to the Wansbeck Rivers and the River Blyth. The original headwaters of these three rivers were captured by the North Tyne, a subsequent river of the Tyne, as it retreated headwards along the Scremerston Coal Series. These rocks were less resistant to erosion than the surrounding rocks.

Fig. 5.26 River capture by the North Tyne of headwaters of Wansbeck and Blyth

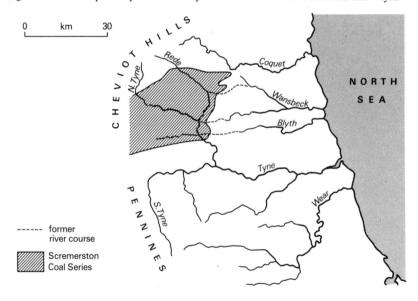

1. *The River Rheidol and the River Ystwyth shown in Fig. 5.27 have captured the headwaters of the Teifi in successive phases of capture. Gap A and gap B mark the probable line of the former courses of the Teifi. Draw two sketch maps to show the course of the rivers: (i) before either capture had occurred, and (ii) after the capture which formed gap A. Comment on the changes in drainage basin areas in these and the present situation.*

Fig. 5.27 The Rivers Rheidol, Ystwyth and Teifi, mid Wales

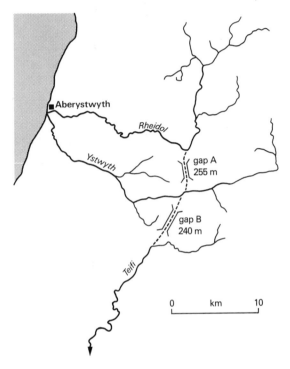

2. *Describe the shapes of the long profiles which you would expect to result from these episodes of capture. The approximate heights of the gaps are given in Fig. 5.27. Illustrate your answer with sketches of the long profiles.*

Key Ideas

A. *Long profiles*
1. The *concave* long profile characteristic of most rivers is a result of the increase in *efficiency* of the channel downstream, enabling lower angled slopes to maintain river flow.
2. *Convexities* in the long profile are the sites of waterfalls and rapids. They mark points where changes in lithology decrease the efficiency of the channel.
3. The channel constantly adjusts its slope, by erosion or deposition, to attain a *graded* state.

4. *Pools* and *riffles* mark areas of *energy accumulation* and *dissipation* in relation to the average energy gradient of the river.

B. *Transverse profiles*
1. The steepness of valley slopes is determined by the relationship between rates of slope movement and river erosion.

C. *Rivers in plan*
1. Rivers can be straight or sinuous. Regularly sinuous sections are called *meanders*.
2. Meanders are like waves and can be analysed as such.
3. Wavelength, width, radius and sinuosity are related to each other and to the discharge.
4. *Helicoidal* flow determines the erosion/deposition pattern in meanders and hence their migration laterally and downstream.

D. *Flood plain formation*
1. The flood plain is built by *point bar* deposition and *overbank* deposition in times of flood.
2. The sudden increase in the width of the river as it overtops its banks causes a fall in velocity and deposition of *levees*.
3. *Braiding* increases the wetted perimeter of the channel and enables more *bed load* to be transported.
4. *Terraces* are abandoned flood plains produced in a variety of ways by changes in the nature of the channel.
5. *Base level* is the lowest level to which a river can lower its bed.
6. A fall in base level increases the potential energy of the river and gives rise to the features of *rejuvenation*, which include knick points, terraces and incised and ingrown meanders.
7. Delta formation occurs where there is a sudden change in velocity as the river enters a standing body of water.

E. *Evolution of drainage patterns*
1. Drainage patterns are a function of structure, lithology and time. They frequently reflect the history of the landscape.
2. *Superimposition* and *antecedence* are ways in which the present land surface inherits patterns generated under previous sets of geological conditions.
3. Headward extension causes *river capture* and enlarges drainage basins.

Additional Activities

1. Survey the long profile of a stream bed using the methods described on page 29. Record also the water depth and velocity (page 115) along your profile. Analyse the data by examining the relationship between bed slope, water slope and velocity.
2. On a meandering section of river, measure the channel width, radius, length and wavelength. Compare these values to those on page 128. Explain any differences you have observed. Many meander studies have made use of experiments conducted in sand trays. It is possible to generate meanders by adjusting the water flow and slope of the surface. The rela-

tionship in these models should exhibit the same relationship as those in real meanders. Braided channels are also easy to generate and it is possible to compare the conditions which produce braiding and meandering.

3. How far do variations in discharge go to explain the fluvial landforms found in river valleys?

4. 'It is impossible to explain present day landforms without recourse to the past.' Write an essay using selected examples of fluvial landforms which reject or support this contention.

5. A glaciated valley about 8 km long and 1 km wide is drained by a river fed by many small tributaries from the valley sides. The floor of the valley has abundant glacial deposits, mainly sand and gravels with interspersed tills (page 188). A band of hard rock forms a rock bar at the lower end of the valley behind which a small lake, about 1 km long, is held up. The river feeds this lake at its upper end, while the outflow from the lake is lowering the rock barrier.

(a) Draw a map to show the main features of the valley.

(b) Describe the fluvial landforms you would expect to find in such a valley and explain how they developed. If downcutting of the outlet continues, what geomorphological developments might you expect in the valley?

Reading

BRIGGS, D.J., *Sediments*, Butterworths, Chapters 3 and 4, 1977

CHORLEY, R.J., *Introduction to fluvial processes*, Methuen, Chapter 7, 1971

*COOKE, R.U. and DOORNKAMP, J.C., *Geomorphology in environmental management*, Oxford University Press, Chapter 4, 1974

*LEOPOLD, L.B. *et al*, *Fluvial processes in geomorphology*, W.H. Freeman, Chapter 7, 1964

HANWELL, J.D. and NEWSON, M.D., *Techniques in physical geography*, Macmillan, Chapter 6, 1973

SPARKS, B., *Rocks and relief*, Longman, 1973

STRAHLER, A.N., *Physical geography*, Wiley, pages 413–437, 1975

6 Glaciers

A. Introduction

1. Glaciation

Within the closed system of the hydrological cycle, referred to on page 77, some precipitation occurs in the solid state as snow. This amount is relatively very small, accounting for about 2 per cent of the earth's total water. Nevertheless it is sufficient to cover 15 million square kilometres or 10 per cent of the

Fig. 6.1 Past and present glaciation in the northern hemisphere (After Bloom)

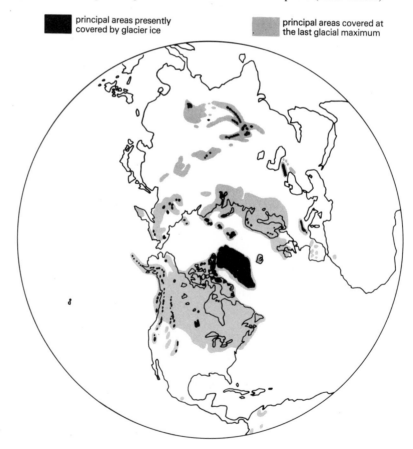

principal areas presently covered by glacier ice

principal areas covered at the last glacial maximum

earth's land area with ice. Antarctica and Greenland together account for 96 per cent of the total ice-covered area, the remainder being found in mountainous country on the other continents.

Many of today's glaciers, ice caps and ice sheets are, however, the shrunken remnants of the ice which covered large areas of the earth's land surface at various times during the last two million years. Geologists refer to this time period as the Pleistocene, when temperatures appear to have fluctuated by about 5–6 degrees C between the warmer interglacial periods and the cooler glacial periods (page 202). During the glacial periods, ice advanced to cover, at its maximum, about 30 per cent of the land area (Fig. 6.1).

2. Snowlines and snow fields

A glacier is formed from a *permanent snow field*. This lies above the *permanent snowline*, the line above which the winter fall of snow exceeds that which is melted during the summer.

The snowline varies seasonally and latitudinally. The winter snowline of course may be considerably lower than the permanent snow line. It is also lower on north facing slopes which receive less direct sun and are in shade for longer periods, if not all the time, as is the case in some very deep Alpine valleys. The snowline is at about 6000 m at the equator, and is at about 3000 m in the Swiss Alps, depending on aspect; as one proceeds polewards, so the snowline generally becomes lower until it reaches sea level in polar regions (Fig. 6.2).

Fig. 6.2 Variations of the snowline with altitude (After Paschinger and Sellers in Sugden and John, *Glaciers and Landscapes*, Arnold, 1976)

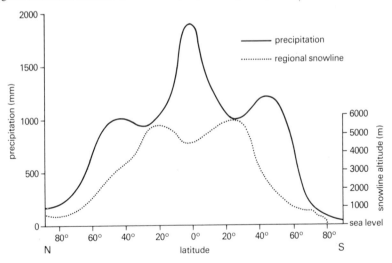

3. Conversion of snow to ice

Freshly fallen snow entraps much air and has a low specific gravity of around 0.1. Such snow occupies ten times the volume of an equivalent amount of liquid water. As the snow is buried by further snow fall the process of compaction occurs and the conversion of snow to ice begins. The specific gravity rises as air is expelled and when 0.55 is reached the snow becomes *firn* (German) or *névé* (French). Firn is the name that is given to snow that has survived summer melting. As successive annual layers accumulate, the firn is further compacted and its specific gravity increases (Fig. 6.3). Meltwater may percolate through the air passages between ice crystals. If this should refreeze, the ice as a whole becomes more dense, since there is less air space. The air is expelled or enclosed as bubbles in the ice. It is clear, therefore, that such a process cannot take place below the permanent snow line.

Fig. 6.3 The increase in specific gravity with depth at Byrd Station, Antarctica (After Mellor)

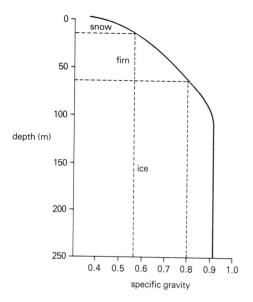

Over many years the firn changes into ice, with a specific gravity of 0.8. Swiss glacier studies have shown that 25 to 40 years are necessary before firn becomes ice and it has been estimated that in Greenland and Antarctica as many as 150 to 200 years are needed to convert firn to ice. Figure 6.3 shows that at Byrd Station in Antarctica ice forms at 65 m depth, the time for this conversion being calculated at 200 years. The colder the climate, the longer is the time required for firn to be converted to ice. Meltwater speeds up the process so ice is formed more rapidly in areas of wet snow than dry snow. For example, glaciers in Alaska require only three to five years for the transforma-tion to ice and this is within a depth of 13 m. Other differences exist between temperate and polar glaciers and these are mentioned later in section C.

For glacier ice of specific gravity 0.9 to occur, further deformation, melting and refreezing is needed to reduce the bubbles of air in size. As this happens, the ice crystals grow in size but hard, blue ice with very small air bubbles will only occur towards the end of a glacier near its snout, as a result of extra pressures from the moving ice. In the Mer de Glace, a glacier in the French Alps, a specific gravity of 0.88 only increased to 0.91 after 50 years of glacier ice flow.

4. Glacier classification

It is usual to classify the various glacier ice masses into:

(i) *Cirque glaciers*, which are small ice masses accumulating in and flowing from rock hollows enlarged by ice. Such glaciers may be quite short in length and confined to part or the whole of the hollow, the cirque. On other occasions, if they are well filled with ice and snow, they feed a valley glacier. The formation of cirques is dealt with at length in section A in the next chapter.

(ii) *Valley glaciers*, which are tongues of moving ice flowing from highland areas where the ice accumulates with time. They are enclosed between valley walls, often the route of a preglacial river valley which is being deepened by erosion by the ice (see Chapter 7A). Plate 6.1 shows a valley glacier, the Flatisen glacier, flowing from the Svartisen accumulation area in northern Norway.

Plate 6.1 Flatisen glacier, Svartisen, Norway (Author's photograph)

(iii) *Piedmont glaciers*, which are enormous lobes of ice formed by the coalescence of valley glaciers as they extend on to the plains beyond the limits of mountain ranges.

(iv) *Ice caps or ice sheets*. Ice caps are large areas of accumulating ice in mountainous plateaux and ranges. The ice flows outwards, sometimes without regard to the underlying relief, to disgorge into surrounding lowland areas via valley glaciers. During the Pleistocene vast ice sheets covered many lowland areas in northern Eurasia and North America (Fig. 6.1). Separate 'streams' of ice flowed within these sheets, even across the sea floor. In many localities of the Antarctic coastline ice shelves form when ice sheets flow out beyond the land and float on the sea. The Ross ice shelf is one example; it is about 500 km wide and the same distance long at its maximum. Ice shelves comprise about 30 per cent of the length of the Antarctic coastline. Today only two ice sheets exist, Greenland (11 per cent of the world's ice) and Antarctica (85 per cent).

ASSIGNMENTS

1. *With reference to Fig. 6.2, suggest reasons for the differences in the height of the snowline, latitude for latitude, in the opposite hemispheres. Consider why the snowline is not at its highest at the equator.*
2. *What factors may cause the height of the snowline to vary from place to place in a restricted locality?*

B. The 'Glacial System'

1. Mass balance studies

It has already been stated (page 151) that a glacier is a tongue of moving ice 'flowing' from a highland area. The upper part of a glacier is known as the *zone of accumulation*, where snow collects by direct precipitation or by avalanching from upper slopes if it should become unstable. This area may be a high plateau where snow may accumulate in sufficient quantity and remain long enough to be converted into firn and then ice. As ice, it flows out and beyond the edges of the plateau and follows routes down preglacial valleys which it

Plate 6.2 Snout of the Österdalsisen glacier, Svartisen, Norway (Author's photograph)

may modify extensively (page 178). Another area where snow may accumulate in sufficient quantity to 'feed' a glacier is a large semicircular basin, a cirque, from which a cirque glacier flows.

The term *ablation* is the opposite of accumulation, and refers to the case of a net loss of ice. Ablation includes loss by surface melting, internal and basal melting, evaporation of surface snow and ice (sublimation), iceberg 'calving' where icebergs break off into the meltwater lakes and streams around the snout of the glacier (Plate 6.2).

The difference between the total accumulation and total ablation for the whole of a glacier over one year is called its *mass balance* or *net balance*. Usually the time interval for measurement is taken from the time of minimum mass in one year to the time of minimum the next year. This *balance year* runs, therefore, from autumn to autumn when summer ablation will have reduced the total ice mass to a minimum. It is interesting to compare this with the 'water year' as discussed in relation to the river basin run-off system in Chapter 4.

To calculate the net balance of a glacier, the difference between the winter balance and the summer balance for the glacier as a whole is calculated. The *winter balance* will be positive (accumulation exceeds ablation) and the *summer balance* negative (ablation exceeds accumulation; a positive net balance indicates the glacier as a whole has gained snow and ice, a negative net balance indicates a loss. If the summer and winter balances are the same numerically, a zero net balance exists (Fig. 6.4).

Fig. 6.4 Glacier balance year: if area on graph is equal (i.e. positive balance = negative balance) there is a zero net balance

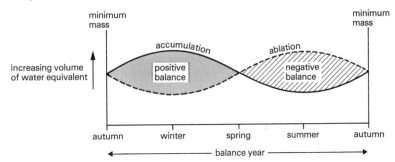

2. The glacier as a 'system'

From the discussion above we can conveniently consider a systems approach to glacial accumulation and ablation. Accumulation represents the 'inputs', the glacier itself the 'storage', and ablation the 'outputs'. Snow (an input) moves through the system as ice and ablates (the output). The period of time for which the glacier (ice in storage) exists depends on its size and the velocity with which it moves. Net balance studies therefore compare total inputs with total outputs.

However, there are wide variations in net balance from place to place on a glacier, but we may distinguish two categories: areas where accumulation exceeds ablation in a balance year (net accumulation area) and areas where ablation exceeds accumulation in a balance year (net ablation area). The boundary between these two areas is called the *equilibrium or firn line* and occurs where ablation equals accumulation in a balance year (Fig. 6.4).

Net ablation is greatest near the snout of the glacier. It decreases to zero at the equilibrium line. Net accumulation is also zero at the equilibrium line (Fig. 6.5) and increases to a maximum above it. These observations refer to conditions throughout the period of a balance year. Clearly accumulation may occur below the equilibrium line on a day when, for example, there is snowfall throughout the whole area of the glacier. Similarly, on a daily basis ablation may occur throughout the whole of the glacier.

Fig. 6.5 Glacier system: if glacier is to remain in equilibrium (i.e. zero net balance) ablation wedge equals accumulation wedge

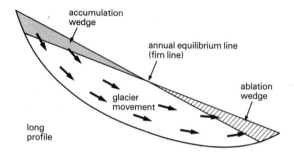

Ice moves continuously across the equilibrium line to balance the volumes of ice on both sides. If the surface of the glacier is to remain in equilibrium, ice must be transferred from the accumulation wedge through the equilibrium line to the ablation wedge under the influence of gravity. Such a condition of equilibrium, where accumulation and ablation are in balance, is described as a glacier in '*steady state*'. Thus, if it is in steady state, a glacier experiencing a large snowfall in its accumulation area will have a steep gradient and rapid flow to transfer quickly the '*accumulation wedge*' to the '*ablation wedge*'. The opposite, of course, applies — small snowfall amounts in the accumulation wedge result in slower glacier velocities and gentler gradients to maintain the steady state.

If steady state does exist at the end of the balance year, the glacier will occupy the same position as the previous year, meaning that the total accumulation equals total ablation over the whole glacier area. If the balance is not achieved, the snout of the glacier will advance or retreat absolutely, depending which of the two, accumulation or ablation, is the greater over the balance year.

Over longer periods of time changes in climates of areas may lead to large scale advances or retreats of glaciers. For example, many glaciers flowing from the Svartisen ice cap in northern Norway show evidence of major advances during the eighteenth century, the 'Little Ice Age' (page 207). Local records suggest the glacial maximum extent to have been between 1750 and 1760, since which time the glaciers have shown steady retreat of their ice margins, leaving behind valleys devoid of soil and vegetation to a very large extent. Plate 6.3 shows a small valley glacier at Svartisen with clear evidence of its having occupied more of the valley in the past. Note the absence of vegetation

154

immediately below the glacier; the eighteenth century advance eroded all the soil and today there is a clear contrast with the adjacent vegetated areas.

Beyond the consideration of net balance studies we must refer to *glacier surges* or *waves* of rapid motion that move through and along the glacier surface up to five times actual ice velocity. They are usually propagated by an abrupt increase in accumulation on the upper part of the glacier, perhaps the result of snow avalanches or unusually heavy seasonal snowfall. The surge, represented by a 'bulge' in the glacier some metres high, travels through the ice and may temporarily lengthen the glacier by pushing the snout downvalley unusually far until the effect of the surge is past. Research seems to suggest that a single variable, such as avalanches, may well be the cause of the surge.

ASSIGNMENTS

1. *Observations have shown that at the snout of a small glacier the equivalent of 600 mm thickness of ice melted in the summer whereas only 200 mm thickness of ice melted at the head of the glacier. Winter snowfall led to 400 mm of accumulation throughout the glacier. Is the glacier in steady state?*

2. *If the equilibrium line were to change up or down the glacier, what effect would this have on the snout of the glacier?*

Fig. 6.6 Glacier response to unusually heavy winter snowfall by 'self regulation'

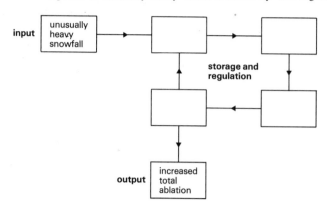

3. *Figure 6.6 represents a simplified glacier system model. Input, storage and output boxes are shown. Assuming we are considering the effects of a winter with an unusually heavy snowfall on the behaviour of the glacier as a whole, insert the following terms in the appropriate spaces: area of ablation zone; total glacier volume; ablation and meltwater production; glacier surface area.*

   Insert + or − along each linkage to show the relationships of adjacent boxes. Identify a negative feedback loop — this shows how a glacier 'damps down' the effects of increased inputs in order to reach a new state of equilibrium.

C. Thermal Classification

One of the most widely used classifications of glaciers is based on the temperature of the ice. From observations it seems that glacier ice temperatures are similar to the mean annual temperature of the air in the areas of ice accumulation. This is the result of ice being a poor conductor of heat, since in its early stages of accumulation it is full of entrapped air and hence each layer of ice keeps the same temperature as that of the air at the time of its accumulation. Although the surface layers may be different depending on the time of the year, when ice is buried at depths below about 10 m the variations average out to correspond to the mean annual air temperature. When glaciers are very thick, over 500 m, as are most of the Antarctic glaciers, the temperature of the ice increases with depth owing to geothermal heat being released through the bedrock on which the glacier rests (Fig. 6.7).

   On this basis we can identify two fundamental types of glaciers:

(i) *Cold base or polar glaciers*

In these latitudes the temperature of snowfall is far below freezing point and as the snow is transformed into ice, the temperature throughout the glacier remains well below freezing point. The ice is frozen to the bedrock on which it is resting and the only meltwater present will be confined to surface streams during the limited periods in the year when the air temperature is above 0 °C, if

Fig. 6.7 Temperature profiles from the first complete boreholes to bedrock in the Greenland and Antarctic ice sheets (After Hansen and Langway (1966) and Ueda and Garfield (1968))

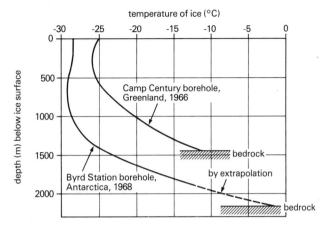

ever. Being frozen to the bedrock affects glacier movement and it is unlikely that polar glaciers can achieve as much erosion as temperate glaciers (see Chapter 7A).

(ii) Warm base or temperate glaciers

This category includes most glaciers today with the exception of the Antarctica and northern Greenland ice caps. The temperature conditions are quite different from polar glaciers, for the temperature throughout, except for the upper surface layer, is near to the pressure-melting point. At atmospheric pressure this is, of course, 0 °C, but as pressure increases with depth through the ice, the melting point is lowered, i.e. water can exist as a liquid at temperatures below 0 °C (Fig. 6.8).

As Fig. 6.8 indicates, only very high pressures will make the pressure-melting point much below 0 °C. In fact, the maximum pressure that has been recorded at the base of the Antarctic ice sheet is about 400 kg/cm², so that even

Fig. 6.8 Effect of increase of pressure on the melting point of ice

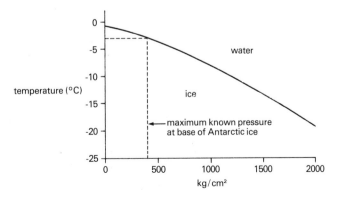

the lowest pressure melting point is only a few degrees below 0 °C, say –2 or –3 °C. Thus, most polar glaciers throughout are well below their pressure-melting point. Interestingly, at the bottom of the Byrd Station borehole in Antarctica (Fig. 6.7) a thin film of water was found suggesting that the ice there is at its pressure-melting point. As it was impossible to take direct temperature readings below 1800 m, extrapolation of the curve on the graph confirmed that the temperature of the basal ice is surprisingly warm, perhaps –2 or –3 °C, compared with surface ice temperatures. Even before this was revealed by the borehole measurements in 1968, several glaciologists had computed theoretical temperature profiles and had suggested that the deepest basal ice in Antarctica might be near to pressure-melting point. Three sources of heat are assumed to create this condition: geothermal, friction with the underlying bedrock if the ice is not frozen to it, and from shearing within the ice mass itself.

In a temperate glacier the ice at the surface in winter will be below 0 °C, yet summer air temperatures will cause melting with meltwater percolating downwards through the ice even to the base. Another factor plays an important part here. This is the latent heat of fusion of ice, which is 3.34×10^5 J/kg. When 1 kg of ice melts, 3.34×10^5 joules of energy in the form of heat are taken from the environment without any change in the temperature of the ice, and conversely, when 1 kg of ice freezes, 3.34×10^5 joules of energy in the form of heat are liberated into the surrounding environment. In a temperate glacier this means that as 1 kg of meltwater from the surface flows downwards to refreeze before it reaches the base of the glacier, enough heat is released to raise the temperature of 6.68×10^5 kg of ice by 1 °C (specific heat of ice is 0.5), raising the temperature of the glacier as a whole. Meltwater can penetrate to the base of a temperate glacier, in vast quantities in summer, especially in the lower valley areas where it is warmer. This film of water at the base usually causes the rate of movement to be faster and hence there is more erosion than in the case of a polar glacier.

D. Glacier Movement

The movement of ice within a glacier is a complicated process and studies over the years, since the sixteenth century, have gradually improved our understanding. However, in the last 40 years more intensive glacier research studies and laboratory experiments have furthered our knowledge of the way in which ice behaves under 'stress'. This situation exists in all glaciers owing to the different pressures exerted by the weight of ice, the contact with the floor and valley sides.

Although we may speak of the 'flow' of a glacier, it is in no way similar to the flow of liquid water in streams and rivers. Water is a viscous substance which reacts to stress conditions in a regular and uniform way; this reaction or rate of yield is known as 'strain'. On the other hand, ice does not behave like a plastic substance which will 'yield' indefinitely at a certain applied stress. Ice 'yields' at an increasing rate as the applied stress increases. Figure 6.9 shows the relationship between these three, and shows how ice eventually behaves plastically as stress conditions increase. In other words, under sufficient stress ice will deform or 'mould' itself with respect to the direction of principal stress.

Fig. 6.9 Graph to show behaviour of ice under stress conditions; ice becomes 'plastic' as the stress rate increases

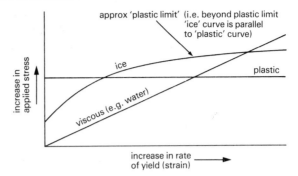

When this occurs the ice is said to have reached the *'plastic limit'*. This characteristic is important as we consider the movement of a glacier.

A glacier's movement can be subdivided into two components, *basal slippage* and *internal flow* (Fig. 6.10).

Fig. 6.10 Movement of the Athabasca glacier, Canada; total surface movement is made up of the two components, basal slippage and internal flow (After Savage and Paterson)

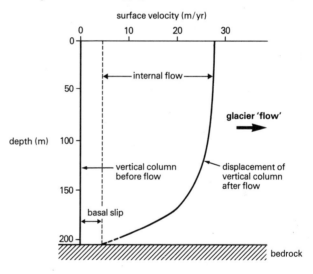

1. Basal slippage

This implies the sliding of temperate glaciers over the bedrock surface (a polar glacier is often frozen to its bedrock and therefore no basal slippage can occur). This process makes possible erosion of the bedrock (page 175). The contribution of basal slippage to total glacier movement varies widely between 20 per cent and 90 per cent. It is at its maximum on relatively steep slopes in summer when the bed is lubricated by meltwater, and at its minimum on flat surfaces or up reversed bedrock slopes. On these latter surfaces glacier movement is caused by thickening of the ice until the centre of gravity of the ice once again falls downhill sufficiently to create movement.

Plate 6.4 The underside of a
Svartisen glacier, Norway
(Author's photograph)

Plate 6.4 shows the underside of the edge of one of the Svartisen glaciers in
Norway as it flows steeply away from the accumulation area. The basal slippage component of its movement can be seen from the shape of the ice which
was moulded around the rock step that lay in its path. The movement shown of
about 1 m was recorded in a period of about six weeks in July–August. Note
the abundance of meltwater in the vicinity which 'lubricates' the base of the
glacier to aid slippage.

Laboratory experiments confirm subglacial observations that basal slippage
occurs by a combination of *regelation slip* (pressure-melting) and *creep*, resulting from stress concentrations around obstacles on the bedrock floor.

(a) Regelation slip

On the 'upglacier' side of the obstacle, pressures in the ice build up sufficiently
to cause pressure-melting locally. The water thus formed trickles around and

Fig. 6.11 Basal slippage by regelation slip; melting occurs on the upglacier side of the
obstacle and refreezing on the other side

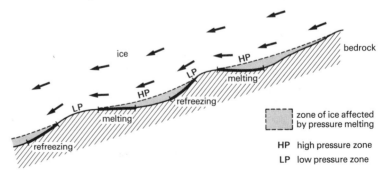

160

over the obstacle where it refreezes in the 'normal' pressure conditions on the 'downglacier' side of the obstacle (Fig. 6.11). This process is maintained by a flow of heat (latent heat of fusion) released on refreezing through the obstacle and surrounding ice to the melting place on the 'upglacier' side. If the ice temperature is below the melting point, this process cannot operate. A further limitation seems to be the size of the obstacle; if it is over 1 m in length, the temperature gradient is too small and the amount of heat conducted is too small.

(b) Creep

As ice encounters an obstacle on the bedrock floor, the stress in the ice is greater than average, and therefore the strain rate will also increase. Thus, the ice becomes more 'plastic' and 'flows' or 'creeps' around the obstacle. The larger the obstacle, the greater the distance over which stress increases and thus creep, as a means of basal slippage, becomes more effective in these circumstances.

2. Internal flow

Internal flow, or internal plastic deformation, refers to movement between or within individual grains of ice as a result of the stresses applied within the ice mass by the force of gravity. Between ice grains, movements occur along *slip planes* where ice grains shear in layers, similar to that of movements within a pack of cards. There may be:

(a) *intergranular movement* where grains slip over one another, as a pile of lead shot would show, or

(b) *melting and recrystallising* of some ice grains under the stress conditions. The most likely cause of internal flow, however, is

(c) *intragranular slip* where, within a single grain of ice, stress causes deformation along parallel planes (Fig. 6.12).

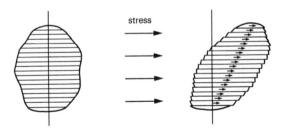

Table 6.1 shows the relative contribution that basal slippage and internal flow make to total glacier movement in different locations. The Meserve glacier in Antarctica is obviously a cold base glacier; such glaciers only flow by internal deformation and perhaps a small amount of fracturing near the surface, where the ice cannot adjust rapidly enough to the stresses within the ice. These ruptures or faults within the ice, when evident at the surface, are called *crevasses* and may extend many metres down into the ice mass (page 164).

Table 6.1 The contribution of basal slippage and internal flow to total glacier movement. The measurements were made from boreholes or tunnels (Source: Paterson, *The Physics of Glaciers*, Pergamon, 1969)

Glacier	Country	Basal slippage (%)	Internal flow (%)	Ice thickness (m)
Aletsch	Switzerland	50	50	137
Tuyuksu	USSR	65	35	52
Salmon	Canada	45	55	495
Athabasca	Canada	75	25	322
Athabasca	Canada	10	90	209
Blue	USA	9	91	26
Skautbre	Norway	9	91	50
Meserve	Antarctica	0	100	80

The rate at which ice may flow varies according to latitude (temperate or polar), the gradient of slope and the time of the year (for temperate glaciers). Thus a temperate glacier in summer flowing down a steep gradient may reach 2

Fig. 6.13 The deformation of vertical boreholes in glaciers:
A Jungfraufirn (M.F. Perutz, 1950): August 1948–October 1949;
B Malaspina glacier (R.P. Sharp, 1953): July 1951–August 1952;
C Salmon glacier (W.H. Matthews, 1959): displacement computed for a one-year period;
D Saskatchewan glacier (M.F. Meier, 1960): August 1952–August 1954.
Note that the vertical and horizontal scales differ in each case. Direction of ice flow left to right. Many of the minor irregularities in the curves for B and C may have no significance, being smaller than the possible errors of measurement (From Embleton and King, *Glacial and Periglacial Geomorphology*, Arnold)

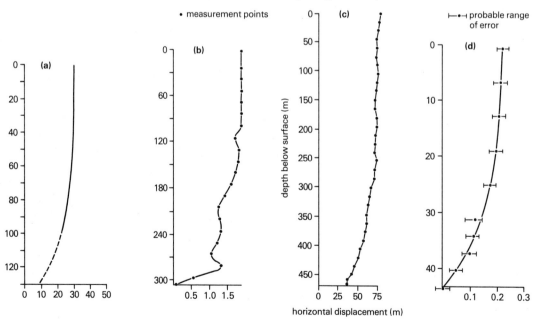

to 3 m per day when the bed is well 'lubricated' by meltwater. On the other hand, a polar glacier moving across a gentle slope may move only a few centimetres in a day. Figure 6.13 shows how great this variation is amongst temperate glaciers: (a) the Jungfraufirn in the Swiss Alps, (b) the Malaspina glacier in Alaska, (c) the Salmon glacier in British Columbia and (d) the Saskatchewan glacier in Alberta.

However, ice margins have been known to advance by up to 100 m or more per day by a surge or wave of rapid motion moving downstream through the glacier (page 155).

ASSIGNMENT

Plates 6.5 and 6.6 both show the Österdalsisen glacier, Svartisen, Norway flowing over hard schistose rocks. In Plate 6.6 the glacier is moving from right to left. Explain the type of ice movement and evidence for it as far as the photographs allow.

Plate 6.5 Österdalsisen glacier, Svartisen, Norway (Author's photograph)

Plate 6.6 Ice caves beneath the Österdalsisen glacier, Svartisen, Norway (Author's photograph)

E. Surface Morphology

1. Crevasses

When a glacier moves down through a valley the frictional drag between ice and rock will create shear stresses that produce crevasses or splitting of the ice in a vertical plane. As mentioned on page 161, sometimes ice is unable to adjust rapidly enough to extra stresses by basal slip or internal flow. Such stresses cause crevasses, often visible on a glacier, the ice splitting apart along a line at right angles to the direction of principal stress (Plate 6.7).

(a) *Marginal crevasses* often occur along the margins of a glacier (Fig. 6.14a), where frictional drag between the ice and the rock of the valley sides causes the margins to move more slowly than the centre. Thus, the crevasses open at an angle to the margin of the glacier. Old ones may be bent or rotated round in the direction of the glacier flow.

(b) *Transverse crevasses* (Fig. 6.14b) are formed when a glacier's gradient steepens and the glacier surface 'extends' itself (*extending flow*) as it 'fits' itself to the steeper profile of the subglacial valley. Since the ice moves more rapidly in the centre than towards the margins, in plan view transverse crevasses are usually convex-shaped upglacier. When the surface of the ice ceases to be extended (*compressive flow*), as when the gradient becomes constant again or reduces, these crevasses close up. Plate 6.8 shows transverse crevasses on the Mer de Glace in the French Alps.

(c) Another type of crevasse pattern can be recognised when a glacier spreads out laterally as the valley widens. Here stresses set up in the ice act across the glacier at right angles to the margins and thus crevasses open up

164

Plate 6.7 Crevasses on the
Mer de Glace, French Alps
(A.S. Freem)

parallel to the glacier margins. Such crevasses are termed *longitudinal crevasses* (Fig. 6.14c).

(d) *Radial crevasses* splay out in the region of the glacier snout (Fig. 6.14d) and as in the case of longitudinal crevasses they result from the spreading out of the ice into a lobe. Once again they open at right angles to the directions of stresses.

Crevasses vary in width from cracks to gaps several metres wide. Often meltwater streams flowing over the ice surface will take a downwards route via a crevasse into deeper parts of the glacier or, if the crevasse is deep enough, may plunge down to reach the rock floor. Studies of polar glaciers suggest that crevasses are often deeper on average than those of temperate glaciers. Crevasses exceeding 36 m have been recorded in the Antarctic and Greenland, whereas temperate glaciers rarely have crevasses deeper than 30 m. This could be explained by the fact that ice forming the top 30 m of a glacier is rigid and brittle. It splits apart in response to stress conditions and crevasses open. Below

Fig. 6.14 Crevasse patterns

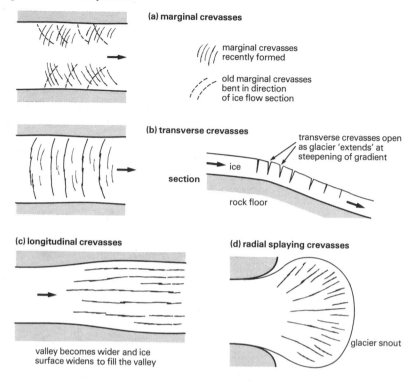

(a) marginal crevasses

//(((marginal crevasses recently formed

;;;; old marginal crevasses bent in direction of ice flow section

(b) transverse crevasses

transverse crevasses open as glacier 'extends' at steepening of gradient

section

→ ice

rock floor

(c) longitudinal crevasses

valley becomes wider and ice surface widens to fill the valley

(d) radial splaying crevasses

glacier snout

Plate 6.8 Crevasses on the Mer de Glace, French Alps (Author's photograph)

166

about 30 m, crevasses once opened are closed rapidly by plastic deformation of the ice. Plastic deformation may even accommodate the extra stresses completely and not allow crevasses to open below this depth at all.

2. Moraines

On its journey from start to finish a glacier may be responsible for transporting vast amounts of rock material or *moraine*. This is the result of the glacier having eroded the rocks of the floor and sides of the valley through which it is passing. In addition, weathering of the rock faces above the glacier by frost shattering supplies a considerable amount of debris carried by rock falls, debris slides and avalanches on to the glacier surface. This moraine on the surface may be recognised either as:

(a) *Lateral moraine* along the edges of the glacier, or

(b) *Medial moraine*, when two lateral moraines have joined where a tributary glacier adds to the main glacier (Fig. 6.15).

(c) *Englacial moraine* is the name for moraine within the glacier itself. Its source may be from the ice–rock interface, where flow within the glacier causes this debris to be raised from the valley floor to higher up within the ice. Alternatively, this moraine may have melted its way into the ice or fallen into a crevasse when one opened in the vicinity of the surface moraine.

(d) *Subglacial moraine* is the debris that is carried along the valley floor by the moving ice. This may have been eroded by the ice (page 175) or it may be englacial moraine that has 'worked' its way down through the ice. Many meltwater streams that flow on the surface become englacial and possibly later subglacial; they transport debris that can add to the morainic load being carried by the glacier. Figure 6.15 shows how subglacial moraine can become englacial moraine when a tributary glacier becomes superimposed on the main glacier.

Fig. 6.15 Moraines of a valley glacier

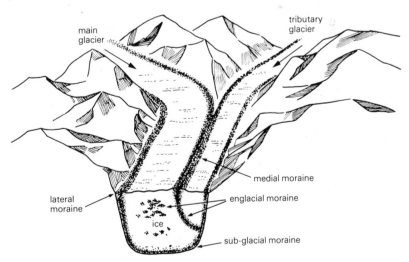

3. Glacier tables

An interesting feature is a glacier table, where a large, usually flat, boulder is supported on a column of ice above the level of the glacier. This is caused by the melting of the ice all around the boulder whilst the ice beneath the boulder has been shaded from the sun and has melted more slowly (Plate 6.9).

Plate 6.9 Glacier tables on a Bernese Oberland glacier, Switzerland (Author's photograph)

ASSIGNMENTS

1. *Plate 6.10 shows the Österdalsisen glacier, Svartisen in the vicinity of its snout. Identify and explain two types of crevasses.*
2. *Identify and explain the patterns of moraines and crevasses on the glacier shown on Plate 6.11. What evidence is there of the source of the material forming some of the moraine?*
3. *What is the evidence of extending and compressive flow as shown on the glacier in Plate 6.8?*

Key Ideas

A. *Introduction*
1. Present day glaciation occupies about a third of the ice and snow cover of the Pleistocene 'Ice Ages'.
2. A lowering of temperatures by only a few degrees was the cause of glacial advances.
3. *Snow fields* lie above the *snowline* where there is permanent snow cover.
4. Snow is converted to *firn* and then to *ice* by *compaction* from the overlying accumulating snow.
5. Glacier ice masses are classified into *cirque glaciers, valley glaciers, piedmont glaciers* and *ice caps* or *ice sheets*.

Plate 6.10 Crevasses on the Österdalsisen snout, Svartisen, Norway (Author's photograph)

B. *The 'glacial system'*
1. The *mass balance* of a glacier is the difference between a year's total *accumulation* and *ablation*.
2. A glacier's *net balance* compares the *positive* winter balance and the *negative* summer balance.
3. The *balance year* runs from the time of *minimum mass*, autumn, to the autumn of the following year.
4. The glacial *'system'* takes account of inputs (accumulation), storage (the glacier itself) and outputs (ablation).
5. The *equilibrium line* occurs on a glacier's surface where ablation equals accumulation in a balance year.
6. A glacier is in *steady state* if total ablation equals total accumulation.
7. Long term climatic changes affect the position of ice margins; *advances* occur if temperatures fall and/or precipitation increases, and *retreats* occur if the opposite happens.
8. *Glacial surges* are waves of more rapid motion than the ice movement, caused by unusually high accumulations in the upper glacier.

169

Plate 6.11 Mer de Glace,
French Alps (A.S. Freem)

C. *Thermal classification*
1. *Polar glaciers*, restricted to Antarctica and north Greenland, are described as *cold base* with temperatures throughout well below freezing point.
2. *Temperate glaciers* or *warm base* glaciers have temperatures throughout near to pressure-melting point. Meltwater in summer may penetrate right to the base.
3. Warm base glaciers generally move faster than cold base glaciers.

D. *Glacier movement*
1. *Basal slippage* is one component of glacier movement. It is accomplished by *regelation slip* — localised pressure-melting — and also *creep*, as ice behaves plastically.
2. *Internal flow*, or *deformation*, also contributes to glacier movement by *intergranular movement, intragranular slip* and *melting and recrystallising of grains* of ice under stress.

E. *Surface morphology*

1. Stresses within the ice cause *crevasses* to open — *marginal, tranverse, longitudinal* and *radial*.
2. Accumulation of debris on the surface of the ice is called *moraine*. Moraines may be described as *lateral, medial, englacial* and *subglacial*.
3. *Glacier tables* show how rates of melting vary in sunlight and shade.

Additional Activities

1. Researchers have suggested that a 'model glacier' provides an interesting investigation into the features and modes of behaviour of glaciers that we have discussed in this chapter. Make a plaster of Paris or fibre glass valley with steep sides, a flat bottom and an uneven and rugged surface; made to scale, this will resemble an actual glaciated valley (page 178). A scale of approximately 1:10 000 works well. The model should be constructed with a corrie at the head of the valley (page 183), and then tilted at an angle of about 5°. Pour a mixture of one part water to two parts kaolin into the corrie. Flow can be measured by inserting matchsticks into the surface. It is also interesting to record how the matchstick bends forwards in the direction of flow of the 'glacier', showing how the surface moves faster than the lower layers. Add fine materials, such as thin filings or powdered charcoal, along the sides of the 'valley' to simulate the accumulation of weathered debris as moraine; if tributary glaciers are made, medial moraines will be formed.

2. Why are geographers so interested in the behaviour of glaciers? Read accounts of various glaciological expeditions throughout the world and find out if the results of research hold anything in common.

3. Draw a landscape sketch of Plate 6.1. Label on it as many features as you can identify that have been discussed in this chapter. To help you appreciate the scale, the thickness of the ice at the snout above the water level is about 10 m.

4. Discuss how the following factors may influence the extent and thickness of snow cover and hence of glacier ice:
latitude, altitude, relief, aspect, distance from nearest ocean, temperature and precipitation.
Build up a systems diagram to illustrate their many linkages.

5. From the photographs used to illustrate this chapter, describe some of the differences between the appearance of the Svartisen glaciers and the Alpine glaciers. Can you think of any explanations?

6. Figure 6.16 is an extract from the 1:25 000 map of the Massif des Écrins near Mt Pelvoux in the French Alps about 50 km south-east of Grenoble. It shows the Glacier de Bonne Pierre and surrounding areas. Study the glacier in conjunction with plates 6.12 and 6.13. Plate 6.12 was taken on the footpath at spot height 2263 m near the snout of the glacier looking east southeast upglacier. Plate 6.13 was taken just below the Col des Écrins at spot height 2980 m looking westwards down the glacier and shows why the glacier is so 'dirty'. The valley sides are composed of a very closely jointed *gneiss* (metamorphosed by the emplacement of a granite batholith nearby).

Plate 6.12 Glacier de Bonne Pierre, French Alps, looking eastwards (A.S. Freem)

Plate 6.13 Glacier de Bonne Pierre, French Alps, looking westwards (A.S. Freem)

Fig. 6.16 (*opposite*) Glacier de Bonne Pierre (© IGN — Paris 1981, autorisation no. 99, carte des Ecrins-Meije-Pelvoux au 1/25 000)

This is weathering easily and is the source of the extensive screes being supplied from the northern slopes of the valley.

(a) Make a tracing overlay of the glacier and the containing valley. Identify and label accordingly prominent glacial features including: accumulation zone, ablation zone, crevasse patterns, moraines and the source of their material, meltwater streams, cirque glaciers and tributary glaciers.

(b) Is there anything to suggest changes in the position of the snout of the glacier?

(c) What does the area around the snout of the glacier indicate about the effectiveness of glacial erosion and transport?

(d) The permanent snow line on a glacier can be called the firn line. Identify its height on the glacier (remember the map shows permanent snow areas in white). How does this compare with the regional snow line as shown in Fig. 6.2?

Reading

BLOOM, A.L., *The surface of the earth*, Prentice Hall, pages 126–137, 1973

STRAHLER, A.N., *Physical geography*, Wiley, pages 523–541, 1973

PITTY, A.F., *Introduction to geomorphology*, Methuen, pages 121–125, 1975

*RICE, R.J., *Fundamentals of geomorphology*, Longman, pages 227–238, 1977

*EMBLETON, C. and KING, C.A.M., *Glacial geomorphology*, Edward Arnold, 1975

*SUGDEN, D.E. and JOHN, B.S., *Glaciers and landscape*, Edward Arnold, pages 1–148, 1976

7 Glacial processes and landforms

As we have seen, the action of ice moving over a landscape is to modify it by erosion and removal of both loose and solid rock and by the eventual deposition of its load. However, the actual mechanisms of erosion are not very well understood, mainly for two reasons. First, it is impossible to know what the preglacial landscape was like in detail and hence it is difficult to assess how much change has resulted. Many glaciated landscapes of the past were subjected to several glacial episodes with different processes at work in the warmer interglacial periods. Also, in the British Isles it is about 20 000 years since the last of the Pleistocene glaciers disappeared and therefore the action of nonglacial processes (fluvial, slopes and weathering) has altered the glaciated scenery in areas such as Snowdonia, the Lake District or the Highlands of Scotland. Second, there is the very practical problem of it being very difficult to observe the mechanisms of glacial erosion at work beneath great thicknesses of ice.

A. Glacial Erosion

Here we are concerned with what happens at the ice–rock interface. Stationary ice has little if any erosive effect. Ice that moves may have a considerable effect in altering the features of the surface, either unconsolidated or solid rock, over which it passes, particularly if 'armed' with loose rock. Ice which is free of such debris has little erosive effect on resistant, poorly jointed rock, though on softer and well-jointed rock, ice movement results in fragments of rock being eroded and transported.

1. Processes of glacial erosion

There are some similarities between ice and water in terms of erosion. Both are capable of exerting great pressures against obstructions in their path and breaking off fragments as they flow past. Both carry rock fragments as abrasive tools, though water has the advantage of greater velocity and turbulence. Ice, however, has the property of greater rigidity and the ability to melt and refreeze as it moves around obstacles in its path that would resist erosion.

(a) Abrasion

A rock fragment at the base of a glacier cannot easily be pressed upwards into the ice and so, if in contact with a rock surface, it will scratch it as it passes

Plate 7.1 Ice eroded
striations (Author's
photograph)

over, marking the rock surface. These markings are called *striations* if very fine
and shallow, or *grooves* if larger, and are good indicators of past glacial
movement (Plate 7.1). As the cutting edge of the fragment of debris becomes
blunted, the rock may be rolled over to expose a new edge and the resulting
striations or grooves may well be deeper but still parallel to the original ones,
though probably offset to one side. Sometimes striations or grooves are discontinuous but rhythmic in occurrence. They are then called *chatter-marks* which
seem to result from the sort of vibratory motion of a cutting tool being forced
across a hard surface. Striations vary in length from 1 cm to 1 m and have been
found on a variety of rock surfaces. Commonly they occur on the sloping rock
surfaces over which ice flowed. The depth of the striations is usually only a
few millimetres but obviously it depends on the hardness of the rock obstacle,
the hardness of the cutting fragment and also the length of time that weathering
has had, since the retreat of the ice, to reduce the surface level of the rock. The
clearest examples will be in those areas from where glacier ice has only
recently melted. 'Clean' glacier ice, carrying little or no fragments in its base,
perhaps only silt and sand, succeeds only in 'polishing' the rock surface.

The product of abrasion is usually very fine rock material called *rock flour*.
Meltwater streams often transport this many kilometres away from a glacier. It
gives rivers into which such streams discharge a very distinctive greenish-grey
colour as rock flour, being so fine, is transported in suspension even at com-
paratively low velocities (Figure 4.20).

(b) Plucking or quarrying

Resistant bedrock has its upglacier side smoothed and striated whilst the lee
side gives an indication of another major process of glacial erosion — *plucking
or quarrying*. This feature is called a *roche moutonnée* (Fig. 7.1)

Plucking or quarrying involves the removal of much larger fragments of rock
than abrasion effects. Just how this process operates is poorly understood but it
is generally regarded as being related to ice at the pressure-melting point (page
160). Rock fragments may be incorporated into the basal ice as the thin layer of

176

Fig. 7.1 Roche moutonnée showing erosive effect of glacier ice; size varies from 1 to 10 m long

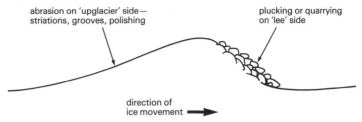

abrasion on 'upglacier' side—
striations, grooves, polishing

plucking or quarrying
on 'lee' side

direction of
ice movement

meltwater that exists at the base of temperate glaciers freezes. This results from a fall in pressure locally, such as on the lee side of a roche moutonnée. Such rock fragments previously will have been loosened or the bedrock will be very soft or perhaps weakened by preglacial weathering. It is highly doubtful whether ice has enough tensile strength to freeze on to fixed obstacles and literally tear them out of the bedrock! Some of the older textbooks have over simplified this process.

It is the opinion of many researchers that plucking or quarrying is the most important erosional process performed by a glacier and is more responsible than any other process for the development of many glacial landforms — cirques, rock steps and roches moutonnées.

Before we leave the topic of glacial erosion to discuss its effects in more detail, however, we must consider some other processes that result in erosion beneath glacier ice.

(c) *Pressure release* (*dilatation*)

The mechanisms of this process have been discussed earlier, in the chapter on rock weathering (page 22). When a glacier erodes, the replacement of a certain volume of rock by ice, of one-third its density, must cause dilatation and separation of the rock along sheet joints. Pressure release is important at or beyond the snout of a glacier or the margins of an ice sheet where the ice front is advancing and retreating seasonally, but will occur anywhere on the valley floor wherever the weight of ice is locally reduced. On the downglacier sides of roches moutonnées or rock steps in the valley release of pressure of the ice may weaken the rock by joints developing parallel to the surface. Weakening of rock in these ways will enable other erosive processes to operate.

(d) *Subglacial water erosion*

Temperate glaciers, in summer especially, have many meltwater streams on their surface, which plunge down crevasses into the glacier (page 165). Many such streams descend right to the valley floor, where they are another cause of erosion of the rock surface. This is particularly true of those which carry a large load of sediment, rock flour or material of morainic origin.

From the above discussion we can understand the complex interrelationships that exist in the consideration of erosion beneath glacier ice. The combination of ice, rock debris and water is an effective agent of erosion and, although in detail our understanding is limited, the effectiveness of this work is clearly visible in many landscapes that have been glaciated.

2. Landforms produced by glacial erosion

As ice flows it modifies the landforms around or over which it passes, creating a more efficient landscape for the future disposal of ice. Thus we must relate the processes at work and the landforms they produce. Much of the effect of ice depends on the type of ice mass involved (page 151); the erosional features produced by the 'scouring' action of ice sheets are less dramatic and often much more extensive than those produced by a valley glacier whose erosive effects were confined to the rock channel or valley through which it flowed. The nature of landforms also depends on the character and shape of the rock surface over which the ice flowed, and the length of time the glaciation lasted. These can be usefully illustrated from a highland region of Britain, the English Lake District. The 1:25 000 Ordnance Survey Leisure Maps of the area are recommended for a detailed study. The Great Langdale Valley provides a good illustration.

(a) Valley cross profiles

The flat-floored and steep-sided valley of today, known as a *glacial trough*, was eroded by a valley glacier as it flowed from west to east (Plate 7.2). As the

Plate 7.2 Langdale Valley looking westwards towards Raven Crag and the valley of Stickle Beck (J.M. Lawley)

glacier moved downslope, it eroded the sides of the valley as well as the floor, particularly the spurs of rock that protruded into the valley and thus into the path of the glacier. This left *truncated spurs*, which give the greatest steepness to the sides of a glacial trough. On Fig. 7.2, find Raven Crag below Harrison Stickle in the Langdale Pikes; this is a good example of truncation of the valley sides. By way of contrast, Plate 7.3 shows the spectacular Lauterbrunnen Valley in the Bernese Oberland, Switzerland. This is a much deeper glacial trough than the Great Langdale Valley, with practically vertical sides where spurs have been truncated.

Fig. 7.2 Great Langdale Valley: plan and long profile

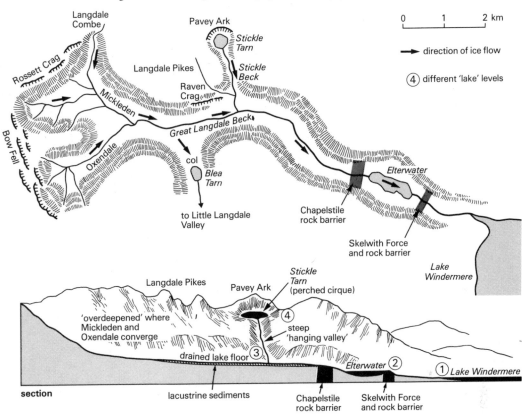

The valley sides of glacial troughs are less steep today towards the valley floor than when ice was present, as weathering and mass movement from the upper slopes has caused scree to collect at their bases. In some literature the cross-profile of a glacial trough is described as 'U' shaped, but this is an exaggerated description of the steepness of the valley sides in all but a few cases, for example the Lauterbrunnen Valley. Other writers have found that the shape approximates more closely to a parabola, taking the form of the curve $y = ax^b$. Each half of the trough is considered separately:

   y = the vertical distance of any point above the valley floor;

   x = the horizontal distance of the point from the origin in the centre of the valley;

a = the coefficient which describes the valley side steepness;

b = the exponent which equals 2 if we consider the shape to be a true parabola (Fig. 7.3).

Two points, say P and Q, are identified on the valley side. The vertical and horizontal distances are measured and at each point (x_1, y_1), (x_2, y_2) the equation $y = ax^2$ is solved, assuming the shape corresponds to a true parabola, and values for a are found. The parabola is drawn using the mean value for a.

For example, let $x_1 = 220$ m, $y_1 = 160$ m, $x_2 = 310$ m, $y_2 = 340$ m

Therefore, $y_1 = ax_1^2$, i.e. $160 = a \times 220^2$, i.e. $a = \frac{160}{220^2} = 0.0033$

$$y_2 = ax_2{}^2, \text{ i.e. } 340 = a \times 310^2, \text{ i.e. } a = \frac{340}{310^2} = 0.0035$$

Therefore the mean value of a = 0.0034.

Fig. 7.3 Glacial trough and related parabola

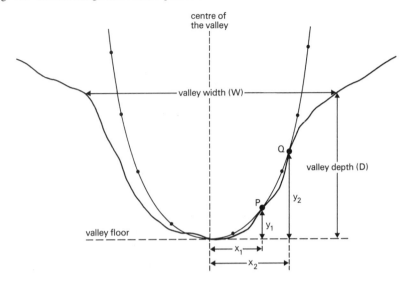

This describes in numerical terms the steepness of the valley sides and is a way of comparing one glacial trough with another.

Since a parabola is an endless curve, it is necessary to define the limits of the cross-section. This can be done by using a *form ratio* (FR) in which FR = D/W where D = valley depth in metres, and W = width of valley at the top in metres.

With reference to Fig. 7.3, suppose the valley depth (D) = 500 m and the valley width (W) = 900 m:

$$\text{Then FR} = \frac{500}{900} = 0.55.$$

A deep glacial trough (e.g. Lauterbrunnen Valley, Plate 7.3) would have a high value of form ratio since the depth is greater than the width. Again this provides a numerical way of describing and hence comparing glacial troughs.

(b) Valley long profiles

The long profile of a glacial trough is often referred to as 'overdeepened'. This describes the very steep gradients found near or at the head of the valley. Towards the mouth of the valley a much gentler slope, sometimes even a reverse slope, occurs. A reverse slope is one which steepens upwards in the downvalley direction, and in a glacial trough is caused by a reduction in the amount of erosion towards the snout of glacier owing to the ice being practically stationary. Soundings in some Norwegian fjords have revealed this reverse slope. Fjords are glacial troughs which have been submerged beneath present day sea level (Fig. 9.38)..

Again the Great Langdale Valley provides an interesting example (Fig. 7.2). Ice accumulated in the upper areas above the Mickleden and Oxendale valleys. As it flowed eastwards the glacier was joined by more ice from the Langdale Combe, a perched cirque (page 183). Almost at its maximum thickness here, it eroded deeply, moving away down Great Langdale Valley with further ice being added from the perched cirque, Stickle Tarn, below Pavey Ark.

The glacier from below Pavey Ark, being smaller than the main Langdale Valley glacier, eroded a shallower trough. However, since the ice surfaces of both glaciers would have been at a common level, the rock floor of the tributary glacier was far above the floor of the main valley. This *hanging valley*, a distinctive feature of glaciated highlands, is now occupied by Stickle Beck, a small stream which descends quickly over waterfalls and rapids to join Great Langdale Beck. Stickle Beck occupies the steep valley in the right middle-ground of Plate 7.2.

It seems some ice was forced up over a low col, spreading southwards to flow into the valley occupied now by Blea Tarn and hence on to the Little Langdale Valley further south. The height of this *diffluence col* is over 100 m above the floor of the Great Langdale Valley. This gives us some indication of the depth of ice that must have occupied the valley — considerably greater than 100 m. It is thought that the upper layers 'sheared off' over the diffluence col and moved away southwards.

Looking at the long profile of the Great Langdale Valley, four different levels can be identified. These levels are shown on Fig. 7.2, and Plate 7.4,

Plate 7.4 Past and present lake levels in the Great Langdale Valley (A.S. Freem)

taken with Stickle Tarn in the foreground, shows all four, with Elterwater just visible and Lake Windermere in the background. In part, these are due to the rock barriers of Chapelstile and Skelwith producing steps in the long profile, which although eroded by the glacier were not removed. Consequently, although deeply entrenched today by Great Langdale Beck, they held back lakes within the Valley. The outflowing river over the Skelwith barrier lowered its level, draining off some of the lake water and reducing it in area. Elterwater is what remains of it today. The long *ribbon lake* that was ponded up behind the Chapelstile rock barrier has been drained entirely, due to its outlet being lowered. However, the flat valley floor covered with lacustrine sediments is clear evidence of its existence in the past.

(c) Cirques and arêtes

Stickle Tarn is a fine example of a *perched cirque or corrie*, a large elevated rock hollow in which ice accumulated and fed a small tributary glacier which joined the main Langdale glacier. Today, this overdeepened hollow is occupied by the tarn. A low concrete retaining dam has been built to increase the tarn capacity for water storage and supply to the surrounding area (Plate 7.4).

Cirques or corries are common features in areas of glacial erosion, and often glaciers flowing from them have contributed much to the impression of dissection in a glaciated landscape. A cirque forms in a shaded or sheltered hollow, often facing in a northerly direction away from the sun, or at a north-facing preglacial valley head, which encourages snow accumulation. Once such a hollow is filled with snow, the periglacial process of *nivation* begins — the enlargement of the hollow by weathering. Freeze-thaw action loosens the rock and snow meltwater washes the weathered material away. Snow lasting more than one year is gradually compressed into ice, and in time the weight of the ice becomes sufficient for a sliding movement to occur. As the hollow is deepened now also by the moving ice, the base of the ice abrades the cirque floor, and the back of the ice plucks at the steepening headwall of the cirque (Fig. 7.4). As the rock is eroded, pressure release probably causes the rock to dilate and sheet joints weaken the rock, hence enabling the erosion to be more effective.

Fig. 7.4 Cirque formation

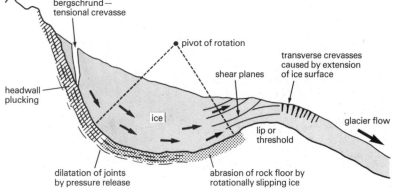

Early twentieth century workers paid special attention to the bergschrund, a deep, tensional crevasse often running around the upper ice surface of the cirque glacier. They suggested that freeze-thaw activity at the base of the bergschrund where it intersected the headwall of the cirque was an important deepening process associated with abrasion by the moving ice. However, later research has shown very small temperature fluctuations in the bergschrund and casts doubts on the occurrence of freeze-thaw at all. Undoubtedly it occurs on the rockwall above a cirque glacier, but it extends beneath the ice surface only to a shallow depth, below which temperatures are constant and seldom fluctuate above 0 °C, even in summer.

W.V. Lewis proposed that ice 'rotates' or 'pivots' on an imaginary point above the cirque. This was based on an analogy with landslides which often show rotational slipping (page 55). This process is associated with the upthrusting of material at the toe of the slip. It is worth bearing this in mind when considering the shape of a cirque which, classically, is overdeepened at the base of the headwall but eroded very little at the 'lip' or 'threshold' of the cirque away from the headwall (Fig. 7.4).

Adjacent cirques and glaciated troughs commonly intersect in cols, but sometimes impressive 'knife-edged' ridges, or *arêtes*, separate them. There is an example of this in the Lake District to the east of Helvellyn; Striding Edge (grid reference 3415), an arête, separates the cirque occupied by Red Tarn to the north from Nethermost Cove to the south.

Plate 7.5 Cir Mhor arêtes, Isle of Arran (Author's photograph)

Plate 7.5 shows the arête ridge between the summits of Goat Fell and Cir Mhor on the Isle of Arran in the lower Firth of Clyde. The granite rocks here are hard and have resisted glacial erosion but back to back cirques at the heads of Glen Rosa (to the left in the photograph) and Glen Sannox (to the right) intersect in the arête.

When three or more cirques intersect back to back, the resulting landform is called a pyramidal peak or 'horn', after the Matterhorn which provides the classic example (Plate 7.6).

ASSIGNMENTS

1. *How do the particular variables — the glacier, bedrock, topography, and time — affect the character of glacially eroded landforms?*
2. *Select various examples of glacial troughs from O.S. maps of the Scottish Highlands, the Lake District, North Wales (Snowdonia) or from foreign maps (e.g. Fig. 6.16).*

Plate 7.6 The Matterhorn, Switzerland (Swiss National Tourist Office)

185

(a) *Obtain values of the coefficient of valley side steepness by the method suggested on page 180 and also their form ratios. Discuss the differences from area to area in your answers.*

The following information is obtainable from the 1:25 000 Outdoor Leisure Map of the SW Lake District for a cross-profile in the Great Langdale Valley at Raven Crag, and shows how to collect the necessary statistics for this example.

Height of valley floor above O.D. = 317 feet (spot height at G.R. 285058)

Horizontal distance to 400 feet contour = 1250 feet (G.R. 284062)

Horizontal distance to 700 feet contour = 1800 feet (G.R. 283063)

Hence from Fig. 7.3,

$x_1 = 1250; y_1 = 400 - 317; x_2 = 1800; y_2 = 700 - 317.$

Values for 'a' can be calculated and a mean value found. You will notice that some maps have not yet been metricated and have heights in feet. This does not affect the calculations, provided that the same units are used for x_1, y_1, x_2 and y_2.

(b) *Construct long profiles and identify irregularities in their slope. Can you explain these? Are they, for example, where a tributary glacier added its ice to the main valley glacier, increasing the erosive power of the glacier at that point?*

B. Glacial Transport

Glacial erosion only becomes significant when rock particles are subsequently transported. There is much evidence to suggest that glaciers and ice sheets are capable of transporting enormous volumes of rock debris over considerable distances. However, a glacier can transport more than just the material it has eroded itself. Mechanical weathering of the slopes above the glacier often supplies material that is carried on to the surface of the glacier by mass wasting.

1. Zones of transport

According to its position, debris is transported in three different zones:

(i) *supraglacially* — on the surface of the ice, e.g. lateral moraine (Fig. 6.15).

(ii) *englacially* — within the glacier as the debris eroded from the ice–rock interface below melts its way through, or is transported along flow lines up through the glacier ice (Fig. 7.5).

(iii) *subglacially* — beneath the glacier along the ice–rock interface. Much of this is directly due to glacial erosion of the rock floor over which it flows.

2. Erratics

A study of erratics reveals how effective a glacier or ice sheet can be at transporting blocks and boulders, sometimes of immense size and weight, over many kilometres. An erratic is a glacially transported block which, on melting

Fig. 7.5 Zones where glacial debris is transported

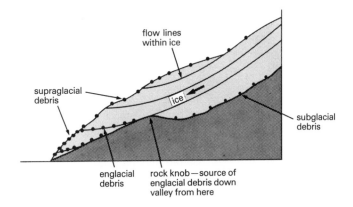

of the ice, is left stranded on rocks of entirely different lithology. Plate 7.7, taken around the Svartisen ice cap in Norway, shows an erratic of igneous rock resting on top of a hard dolomitic limestone. It is interesting to note how it has protected the underlying limestone from solution (page 25) whilst the limestone surface elsewhere has been lowered; this is analogous to the formation of glacier tables (page 168).

Often the erratic can be traced back to its original source outcrop and provides useful information on the course followed by the ice, especially when ice sheets are responsible for its transport. For example, erratics from the island of Ailsa Craig off the Ayrshire coast in the Firth of Clyde have been found on the South-west Lancashire Plain, proving the direction of ice flow and a journey of at least 240 km across what is today the sea. The particular type of granite of which these erratics are composed is unique to Ailsa Craig and thus provides indisputable evidence of their origins. The same granite has been found as far south as St Davids in north Pembroke as erratics, a journey of about 400 km. Some erratics are quite enormous; a chalk erratic in Norfolk is 800 m long! A quartzite erratic in Alberta, Canada, is 24.5 m × 12 m × 9 m and weighs over 18 000 tonnes.

Plate 7.7 Glacial erratic, Svartisen, Norway (Author's photograph)

C. Glacial Deposition

If the effects of glacial erosion are most obvious in highland areas, it is in lowland areas that glacial deposition is most extensive.

1. Processes of glacial deposition

Wherever the ice melted, either in the form of a glacier or as an ice sheet, large quantities of debris were deposited on the landscape. This happened particularly around the snout of a glacier or around the margins of an ice sheet, changing the form of the preglacial landscape.

Before looking at particular landforms resulting from glacial deposition, it is worth noting the threefold division of such .deposits, which are collectively called glacial *drift*:

(i) *Glacial till* is unsorted and unstratified debris, stranded over the landscape and deposited by direct ice action. It is composed of fragments of rock of all shapes and sizes, ranging from large boulders to small clay particles, mixed randomly together (Plate 7.8). Often in Britain this till is referred to as *boulder-clay*.

Plate 7.8 Section through boulder-clay (Author's photograph)

(ii) *Ice-contact stratified drift* is a second group in which debris is partly sorted by water action and roughly stratified. This is deposited in the vicinity of melting ice.

(iii) *Outwash deposits* form the third group. They are carried and graded by glacial meltwater and are therefore well sorted, according to particle size, and stratified.

2. Effects of glacial deposition

(a) Drumlins

(i) Upper Ribblesdale, North Yorkshire

An example of the effects of glacial deposition on the landscape is in the upper Ribble Valley, south of Settle in North Yorkshire. There is much evidence to assume that glacier ice flowed down the Ribble Valley from the north (Fig. 7.6). In the area around Hellifield, the ice seems to have divided, some flowing

Fig. 7.6 The Ribble Valley around Hellifield

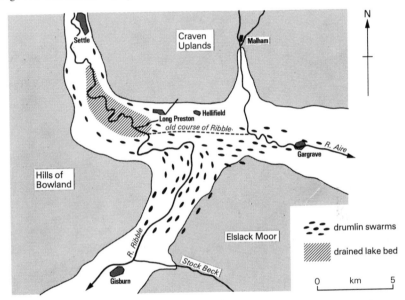

eastwards to the Aire Valley and some flowing southwards into the valley today occupied by the lower Ribble. Here much boulder-clay was deposited by the melting ice, or from ice overloaded with material, to be shaped and 'stream-lined' by later ice flowing over it into features called *drumlins*. These are elong-ated, rounded hills of boulder-clay, asymmetrical in form, with a blunt 'nose' facing the direction from which the ice advanced and a tapering 'tail' facing 'downglacier' (Fig. 7.7). In the Ribble Valley they occur in swarms, each varying in size from a few metres to 30 to 40 m high and from 50 to 500 m long (Plate 7.9). The swarms here have given rise to what is often called a *'basket of eggs'* topography. The long axes of the drumlins are parallel to the direction of ice movement and from Fig. 7.6 it is easy to see the division of the ice along the Ribble and Aire Valleys respectively.

Many geographers today think that the preglacial upper Ribble was a tribu-tary of the Aire and that the valley around Hellifield was blocked by the drumlins left by the Pleistocene glaciers. So when the ice had disappeared completely the upper Ribble found a new and easier route to follow, through and around the drumlins, southwards towards Gisburn, thus giving today's Ribble

Fig. 7.7 Drumlin shape and measurement

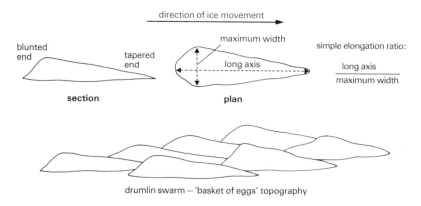

drumlin swarm—'basket of eggs' topography

Plate 7.9 Ribblesdale drumlins (Author's photograph)

Valley route. Also at one stage, probably just after glaciation, the drumlins so effectively blocked the River Ribble that a large lake was ponded up between Settle and Long Preston (Fig. 7.6). Its level rose until the outflow overtopped a low col through the drumlins to allow the river to flow. Since that time erosion of this col has lowered the lake level and today all that remains is a flat, wide valley very different from the valley downstream. The present day river meanders slowly across it in a channel that is embanked artificially to reduce the frequency of flooding (Plate 7.10). (Refer back to Chapter 4E.)

(ii) Drumlin measurement

The shape of drumlins can be measured by using the *elongation ratio*. The length of the drumlin (its long axis) and the maximum width are measured (Fig. 7.7):

$$\text{The elongation ratio} = \frac{\text{length of drumlin}}{\text{maximum width}}$$

190

Plate 7.10 Drained lake bed below Settle in the Ribble Valley (Author's photograph)

For example, suppose a drumlin is 40 m long and 10 m wide at its maximum:

$$\text{The elongation ratio} = \frac{40}{10} = 4.$$

This measurement, together with a compass bearing of the long axis, usefully describes the drumlin.

As with the fitting of a parabola to a glacial trough, so the elongation ratio enables one to compare drumlins in different locations quantitatively rather than to rely on verbal description:

(b) Till fabric analysis

Just as the form of a drumlin as a whole is orientated in relation to ice movement, so often it is revealed by close examination in the field that the pebbles which make up the boulder-clay have also been orientated as they were deposited (or later redeposited by other ice). Till fabric analysis, as such studies are called, may well reveal that the long axes of angular pebbles share an approximate common orientation. A compass bearing measurement can be recorded for a sample number, say 50, of the long axes of pebbles taken from a locality where boulder-clay is exposed. Such exposures must not have been disturbed so that the pebbles remain with their original orientations; it is wise to remove the surface materials and work on the newly exposed pebbles. If these measurements are grouped into classes of say 15°, they can be plotted on circular graph paper as a rose diagram (Fig. 7.8). The length of the direction lines is proportional to the number of pebbles in each class. On the diagram each ring represents four pebbles. When plotting a pebble class pointing north-west (315°) at one end, clearly the pebble class points south-east (135°) at its other end; so both directions are plotted making the rose diagram symmetrical. The points at the ends of the direction lines are joined to complete the diagram.

Fig. 7.8 Till fabric analysis showing pebble orientation of 50 pebbles from Endon till, Endon Valley, to west of Leek, Staffordshire

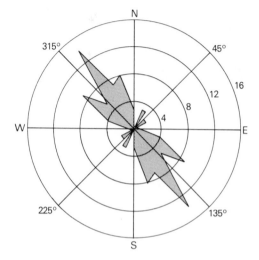

However, the direction of ice movement in the field can only be one of the two directions, and which it is would have to be decided using other information, such as drumlin orientation, sources of glacial erratics found in the area or knowledge of local areas of ice accumulation in the Pleistocene, such as nearby cirques.

Glacial deposits are usually composed of angular pebbles since very little, if any, rounding has taken place by attrition; in fact, rounded pebbles are usually only present when they have been transported by fluvial processes for some time (Chapter 4, page 116).

(c) Moraines

So far we have concentrated upon the glacial deposits stranded under the melting ice and subsequently reshaped by later ice passing over them. Sometimes an advancing glacier or ice sheet is so overloaded by debris that it simply dumps or smears its load on to the valley floor, as a thick mantle of till known as lodgement till. In the later stages of glaciation, when the ice margins are shrinking, much ablation of the ice takes place, stranding its load across the landscape as ablation till (Plate 7.11). Extensive areas of the North European and North American Plains show this hummocky landscape today. In places the underlying relief has been masked completely and preglacial valley systems may be obliterated. Often several advances and retreats of the ice margins

Plate 7.11 Ablation till, Zmutt glacier, Switzerland (A.S. Freem)

create a number of layers of ground moraine, *multiple till units*, which after examination have formed a basis for Pleistocene stratigraphy and chronology.

The maximum extent of the ice sheets or a glacier is marked by a *terminal moraine* which is the sum total of all the debris brought down by the ice to its furthest position. Sometimes the ice pushes this material ahead of it like a bulldozer to form *push moraines*. Otherwise it is the accumulation of supra-glacial, englacial and subglacial moraine (Fig. 7.5). Smaller 'temporary terminal' moraines or *recessional moraines* (Plate 7.12) upvalley of the final terminal moraine may represent periods of 'standstill' of the ice-margin during the retreat stage of the ice. Often stretching right across many glaciated valleys in Britain, these banks of moraine stand out clearly today, perhaps deeply incised by the present day river flowing through them.

Sometimes supraglacial debris is so thick that *ice-cored moraines* are formed. The moraine insulates the ice underneath, preventing it from melting whilst the surrounding ice melts. The ice-core may melt much later, leaving behind a depression in the moraine covered landscape known as a '*kettle*'. *Kettle lakes* may occupy these depressions if they are large and deep enough.

Upvalley of these morainic features, lacustrine deposits often make a flat valley floor, proving that a lake, perhaps of meltwater from the ice, was once dammed up there. Sometimes erosion terraces, with lakeside beach deposits on

Plate 7.12 Recessional moraines, Cirque de Pombie, French Pyrenees (Author's photograph)

them, cut at former lake levels, are further evidence of the existence of a lake.

There are many small scale examples of terminal moraines, but some of the most extensive lie across the North European Plain stretching from Denmark, through Germany, Poland and into the USSR (Fig. 7.9). This map shows how the maximum advance of the last glaciation, which occurred relatively late in the Pleistocene period, about 17 000 years ago, was not as extensive as some of the preceding glaciations. Of the earlier, less extensive glacial advances there is little trace, since when an ice sheet or glacier advances it tends to destroy the evidences of any earlier glaciation, such as the moraines, till and outwash features. A discussion of glacial advances, retreats and warm interglacial periods of the Pleistocene in Britain follows in section F.

ASSIGNMENTS

1. *What can you tell about the direction of ice movement across the landscape shown in Plate 7.9? Are the depositional features the product of ice sheet or glacier action?*
2. *What is the main orientation of the till fabric represented by Fig. 7.8? The results were collected from a site in a valley which trends north-west to*

194

Fig. 7.9 Pleistocene glaciation in Europe

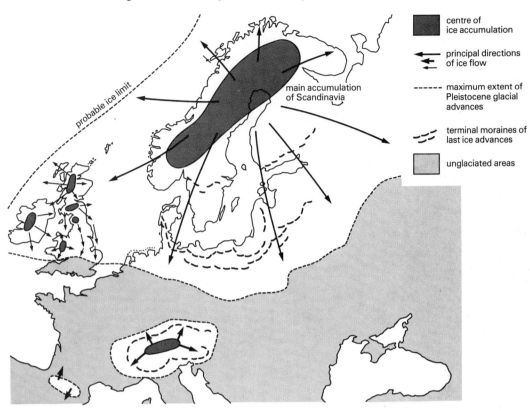

south-east downvalley. Does this enable you to make any further conclusions about the formation of the till?

3. Compare and contrast the form and mode of occurrence of drumlins and moraines. What conditions favour the development of each?

D. Fluvio-Glacial or Meltwater Deposition

1. Outwash plains

Many glacial deposits are laid down initially under water or are later transported and redeposited by meltwater. These conditions create a number of different features collectively known as *outwash features*.

The *outwash plain* which lies below the snout of a glacier or away from the margins of an ice sheet is composed of glacial sands and gravels transported and deposited by meltwater streams. Such deposits are sorted or graded into sizes, the coarsest being found near the ice, with the deposits becoming progressively finer as distance from the ice increases. Sometimes the finest, rock flour, is transported many kilometres before it is deposited (page 176).

The discharge of meltwater streams across an outwash plain varies seasonally and daily. Summer discharges are obviously greater than winter ones, because of the faster rate of melting in the warmer weather. There may be no discharge at all during the winter months. Discharges also fluctuate from a maximum in the late afternoon, when the heat of the day has created much melting and therefore surface run-off of meltwater is high. Such variations in discharge may give rise to a vertical stratification of sediments; coarse sediments are brought into a locality only at times of high discharge, the finer sediments are associated with low·discharges.

Scandinavian researchers have recognised these variations in size as a rhythmic sedimentation associated with lake floor deposite, known as *varves*. Each layer of sediment represents an annual cycle of deposition (Fig. 7.10).

Fig. 7.10 Varve sediments: section through lake floor deposits

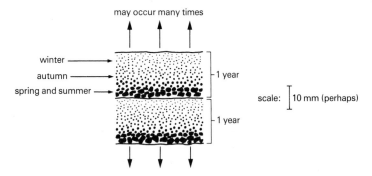

The coarse material is laid down during the spring snow melt. Above this the sediment grade becomes smaller down to very fine material laid down during the winter months. At that time little if any meltwater flows into the lake, allowing the fine suspended material brought in earlier to settle.

Fluvio-glacial sediments are more rounded than the more angular glacial sediments, owing to attrition (page 105) during transport. Measuring pebble roundness (page 117) within glacial tills and in an outwash plain shows their differences in shape.

The gradients of many outwash plains are quite steep, since (page 122) the streams build up their valley floors to maintain flow and thus are able to transport the great volumes of sediment load. When in an enclosed valley, these outwash plains are called *valley trains* and they too are steep, perhaps up to 50 m per kilometre. Frequently streams flow through braided channels (page 134) which reflect the vast amount of sediment, pebbles and boulders that are moved from time to time. Maximum movement of this material takes place in the summer when the greatest discharges occur.

2. Ice-contact stratified drift

Within and under a glacier there are a number of situations in which meltwater deposition produces *ice-contact stratified drift* (Fig. 7.11). All these situations occur in tunnels or at the end of tunnels in the ice. If a stream in a tunnel divides, the discharge in each tunnel reduces, the energy to transport decreases and deposition results (Fig 7.11a). Some tunnels deepen in places and in the deep, perhaps stationary, water, deposition occurs (Fig. 7.11b). Hydrostatic

Fig. 7.11 Situations where ice-contact stratified drift may accumulate within englacial or subglacial streams

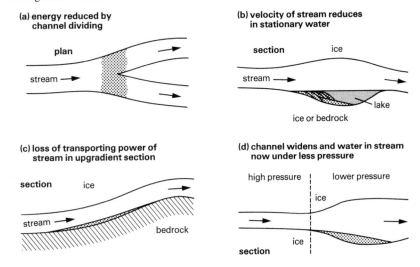

(a) energy reduced by channel dividing

plan

stream →

(b) velocity of stream reduces in stationary water

section   ice

stream →

ice or bedrock

lake

(c) loss of transporting power of stream in upgradient section

section   ice

stream →

bedrock

(d) channel widens and water in stream now under less pressure

high pressure | lower pressure

ice

ice

section

pressure may force a stream to flow upwards for a short way; in the upgradient section the stream loses its power to transport and deposition follows (Fig. 7.11c). Figure 7.11d shows another situation where deposition occurs — as a tunnel widens or deepens, the water under high hydrostatic pressure is now under lower pressure and deposition takes place. This assumes that the water filled the tunnel completely in the constricted section.

However graded or stratified the deposits may be initially, when the ice wastes away, the depositional features formed will probably be much altered by further glacial or fluvial transport or simply by collapsing as the tunnel disappears and leaves the deposit unsupported. Consequently, much mixing of the sediments occurs. This explains why so many fluvio-glacial landforms created in this way are discontinuous, with rapid changes in very short distances of coarse and fine deposits. It also explains why they are difficult to interpret. Three particular features are identified as being formed in this ice-contact environment (Fig. 7.12):

(a) Kames

These are small mound-like hills composed of sands and gravel. They vary in dimension, shape and stratification. The collapse of sediment after the ice melts disturbs or even destroys the stratification. They may be formed in large crevasses where surface streams have washed debris in, stranding it there as the water seeps through the debris and drains away at the bottom of the crevasse. When the ice melts the debris is left as a mound, *a kame*, on the valley floor. Others are formed when streams flowing off the glacier build up a small delta in the static water of a small ice-marginal lake. If such streams emerge into the lake at some height above its level, the kame that results will build up to this height from the lake floor level, provided the ice margin is stationary for long enough. On ice melt, such a kame will be very much altered in shape as the

197

Fig. 7.12 Ice-contact drift features (After Flint)

(a) glacial landscape

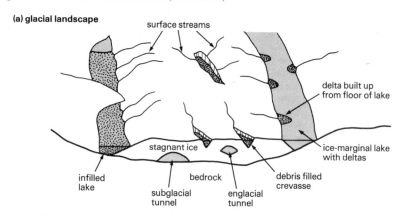

(b) postglacial landscape

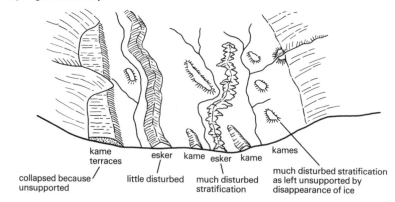

former ice-contact side will be entirely unsupported and collapse to the valley floor with consequent disturbance of the remaining sediments.

(b) Kame-terraces

This is a more continuous depositional bench along the valley side in the space between it and the glacier, again formed by supraglacial and englacial streams discharging their sediment into a lake which becomes filled in time. Once more, after the ice has melted, the ice-contact side of the kame-terrace is unsupported and may slump down, disturbing the sedimentary structures. By mass movement, this side eventually stabilises at the particular angle of rest for the sediment (page 49), whatever it may be.

(c) Eskers

Eskers are long sinuous ridges composed of sand and gravel sometimes branching or joining. They may be 20–30 m high and meander gently across the drift-covered landscape for many kilometres. As Fig. 7.12 shows, they represent deposition of sediments that have occurred in an englacial or subglacial stream tunnel. After melting of the ice, the sides of the esker slump down to a suitable angle of rest, but horizontal bedding of the deposits is usually preserved at the centre of the esker.

It is clear from this discussion that for kames, kame-terraces and eskers to form the ice must not be advancing but must be stagnant and wasting away *in situ*, otherwise these features would never be preserved. Thus, they are only formed in the late stages of glaciation when the climate is warming and glacier or ice sheet margins are in retreat.

These ice-contact drift features are all composed of clean sand and gravel and have often been excavated as sources of material for the construction industry. Thus, few are preserved in the more accessible areas today, but such quarrying operations do sometimes enable one to see the often much disturbed or even chaotic structures that themselves may provide interesting studies of sedimentation.

ASSIGNMENTS

1. *Explain the differences in sedimentary structure that may have resulted between the two eskers shown in Fig. 7.12.*
2. *Describe the features of deposition shown in Plate 7.13 of the Kandersteg Valley in the Bernese Oberland, Switzerland, an area that underwent valley glaciation in the Pleistocene. What can you conclude about the stage of*

Plate 7.13 Glacial deposition in the Kandersteg Valley, Bernese Oberland, Switzerland (Author's photograph)

glaciation in which deposition occurred here? Suggest what has happened to the features during the postglacial period.

3. *What differences would you expect to find in a field study of eskers and kames? Would their sedimentary structures be similar?*

4. *How easy is it to distinguish between a drumlin and an esker?*

E. Glacial Meltwater Effects

In all the studies we have made of glacial erosion and deposition and the associated landforms, we must remember that during the Pleistocene the average temperatures were only about 5 or 6 °C lower than those of today. Particularly during the summers of the glacial period, there would have been much meltwater present around, in, under and proceeding from the glaciers and ice sheets. These meltwaters produced a number of landforms.

1. Overflow channels and spillways

Many high level channels in upland areas today are the result of glacial meltwater flow with consequent fluvial erosion, as lakes entrapped by ice rose in height and perhaps found cols to overflow into adjacent valleys (Fig. 7.13). Many such channels, or spillways, are dry today, but their existence can only be explained with reference to glacial meltwater modification. Possibly the new course of the River Ribble (Fig. 7.6) southwards towards Gisburn was cut by meltwater overflowing from the lake between Settle and Long Preston.

Fig. 7.13 Glacial spillway formation

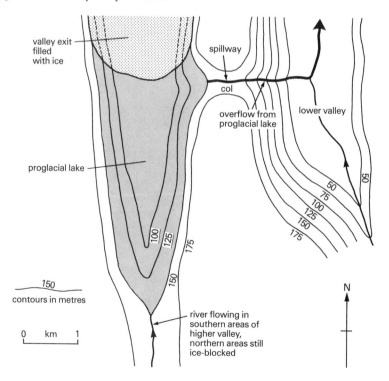

2. Proglacial lakes

Meltwater lakes adjacent to glaciers or ice sheets are called *proglacial lakes*. Researchers have found evidence of these in several areas of the Pennines and around the Cleveland Hills and the Vale of Pickering in Yorkshire (Fig. 7.14). Ice surrounded the Cleveland Hills and several proglacial lakes developed. Lake Eskdale rose to 216 m O.D. before it found a col to spill over into Lake Glaisdale. This was joined to Lake Wheeldale which rose to a maximum height of 203 m O.D. Its overflow channel was to the south, which today is the 5 km long Newtondale Gorge, discharging into Lake Pickering, a large lake between the Cleveland Hills and the Yorkshire Wolds. The village of Pickering today is sited on the gravel delta that was formed where Newtondale overflow water reached the static waters of Lake Pickering and dropped its load. Today Newtondale is almost dry, with just a small stream rising half way along it and flowing southwards into the fertile Vale of Pickering. Lake Pickering rose to a maximum of 68 m O.D. before it spilled southwards into the Vale of York. Today the River Derwent uses this channel, rising as it does in the east of the Vale of Pickering, flowing away from the North Sea into the Vale of York eventually to join the Ouse. The drainage of this area today is a good example of how glaciation has modified a rather different pattern of drainage, for the Derwent formerly flowed direct to the North Sea.

Fig. 7.14 Proglacial lakes and overflow channels

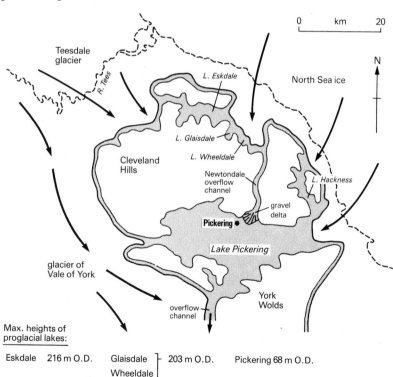

201

3. Subglacial drainage channels

In certain areas, water-eroded channels cut in solid rock are thought to have been formed by subglacial streams. What is rather unusual about some of these is that they have irregular long profiles with a series of ridges and basins. Others have one main crest and they are termed *'humped'* or *'up and down'* channels. Some have been measured as rising 80 m in Lanarkshire, Scotland, whilst one uphill section in Greenland rises 142 m.

These humped channels are most easily explained by water flowing under hydrostatic pressure in subglacial tunnels. This suggestion is perhaps confirmed by the occurrence of ice-moulded sides to the channels.

F. Britain During the Pleistocene Glaciations

The ice which spread out over Britain from the accumulation centres in Scotland, northern England and Wales did so not once but a number of times. Each of these glacial 'pulses' has been named, as shown on Fig. 7.15. The glacial episodes, when temperatures were probably 5–10°C lower than present, were interspersed with periods at least as warm as present. There are at least two of these *interglacial* periods.

Fig. 7.15 Climatic events in Britain during the Pleistocene

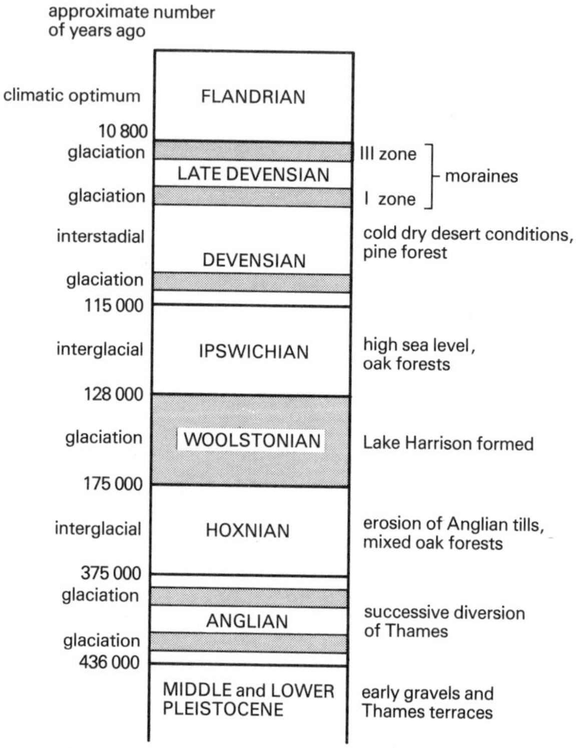

1. The Anglian glaciation

The *Anglian* glaciation is the name given to the earliest of the glacial advances. It occurred about 500 000 years ago and Fig. 7.16 shows it extended southwards over England possibly spreading as far south as the Scilly Isles to the south-west of Lands End. Two major cold periods are recognised within this period when glaciers twice spread as far as the Thames valley.

Fig. 7.16 The extent of ice at various stages of Pleistocene Britain

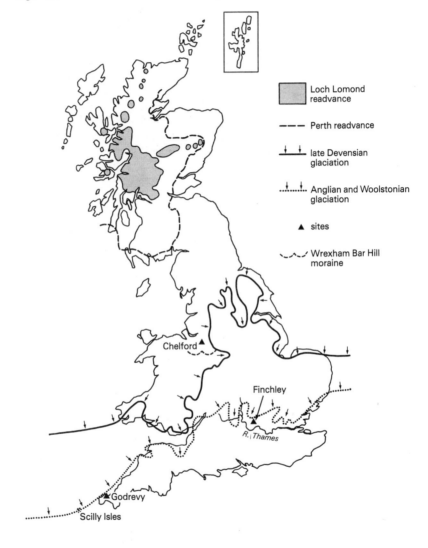

In the middle and lower Pleistocene, before glaciation, the Thames flowed north-east, deposited gravels and formed terraces (page 135). The first phase of the Anglian glaciation formed the St Albans Lobe, blocked this earlier route, (1) in Fig. 7.17, and forced the Thames to take a more southerly route (2) towards Ware. Then a second advance of the ice forming the Finchley Lobe, blocked this valley pushing the Thames into its present course, 3 in Fig. 7.17.

Fig. 7.17 Early courses of the River Thames

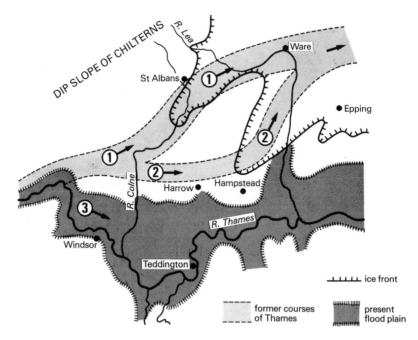

The gravels and terraces left by the Thames when it abandoned its former courses contain rocks (Lower Greensands) which could only have come from the south, adding to the evidence that the river flowed from south-west to north-east. The terraces of the present course of the Thames, described on page 136, were formed in the interglacial which followed the Anglian glaciation.

2. The Hoxnian interglacial

This was a period of relative warmth when a mixed oak forest grew in lowland Britain. The Thames terraces have shown that the straight-tusked elephant (*Elephas antiquus*), the rhinoceros and early horses (*Equus*) lived there. Elsewhere there is evidence of humans from hand axes and fossil skull remains, found in the Boyne Hill Terrace of the Thames.

This period allowed rivers to cut down through the tills deposited by the Anglian glaciers so that these now lie mainly in hill top positions.

3. The Woolstonian glaciation

The extent of the ice during this cold period is unclear. Scattered remnants of boulder-clay of this age lie across the Midlands and East Anglia. In the Worcestershire area, Welsh ice and ice coming from the north, possibly from the Irish sea and Scotland, blocked the valleys of the Avon, Tame and Soar and created Lake Harrison, shown in Fig. 7.18. For almost 10 000 years it lay ponded between the ice fronts and the Jurassic escarpment. Water from the lake flowed over low points in the scarp and into the Thames Basin. Later the ice pushed across this lake bed and sealed the lake deposits beneath a layer of boulder-clay.

Fig. 7.18 Lake Harrison (After Shotton)

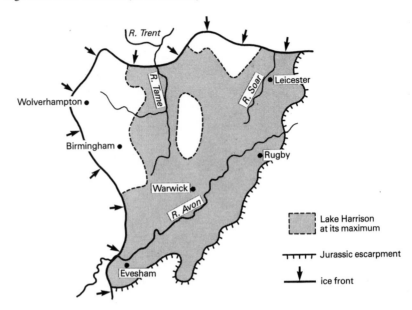

4. The Ipswichian interglacial

This phase was one of a fluctuating climate in which oak forest at times grew in lowland Britain. Excavations in Trafalgar Square revealed traces of elephant (*Elephas primigenius*) and hippopotamus, which spread almost as far north as Leeds at this time.

Figure 7.19 shows the abundance of several tree species during this period. Pollen is very resistant to decay; it can be extracted from peaty deposits and

Fig. 7.19 Simplified pollen diagram of the Ipswichian interglacial (After West)

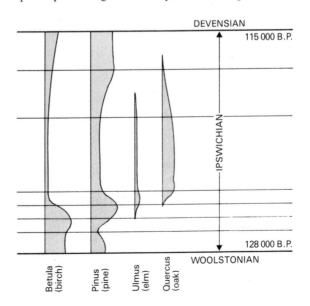

counted to indicate which were the dominant plants at the time. The warming at the start of the interglacial is indicated by fall in the abundance of birch (*Betula*) and pine (*Pinus*), and a corresponding increase in the amount of oak (*Quercus*) pollen. Similar diagrams for the Hoxnian interglacial show much greater proportions of lime (*Tilia*) pollen enabling distinctions between the periods to be made.

Sea level at this time may have been up to 15 m above present sea level, depositing the beach materials shown in Fig. 7.16 at Godrevy Point in Cornwall.

5. The Devensian glaciation

The name Devensian given to the last major glacial episode comes from the 'Deva', the Roman name for Chester. In Highland areas the deposits are almost all of this age. Only at the margins of the ice sheets are older glacial deposits preserved.

There appear to have been two main cold phases in the Devensian. The later one began about 26 000 years ago and reached a peak about 18 000 years ago. This deposited a till in Cheshire (Fig. 7.20) which contains shells picked up from the floor of the Irish Sea. In the Chelford area of Cheshire prior to this ice advance the climate had been cold and dry. Alluvial fans spread quartz sands westwards across the Cheshire Plain from the Pennine foothills. The wind moving sand across the surface of these fans cut faces on pebbles, forming *ventifacts*, some of which have a coating of manganese oxide, or 'desert varnish'. Frost wedge casts found in the sands (page 219) indicate that the ground was subsequently frozen and periglacial conditions existed in advance of the migrating ice sheets.

Fig. 7.20 Section through deposits at Chelford, Cheshire (After Worsley)

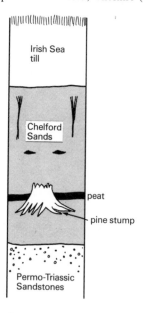

There were, however, warmer periods when pine trees grew. A layer of *in situ* stumps associated with a peat is found within the sands. Such warm periods are referred to as *interstadials* to differentiate them from interglacials, since they were not as warm as the present climate.

The glaciation which covered these sands with a till layer was both rapid and extensive. Figure 7.16 shows that ice pushed south to cover much of Wales and northern England. In Scotland it may have been 2 km thick. Its maximum extent is marked by a feather edge of till rather than a terminal moraine. The Wrexham Bar hill moraine marked on the map is a major feature formed during a halt in the recession of the ice-front from the Cheshire Plain. There are many such recessional moraines on lowlands inside the limit of the Devensian ice.

The ice-front retreated relatively steadily until about 14 500 years ago, when only small cirque glaciers remained in Wales and the Lake District. A slight cooling of the climate caused the glaciers to advance and push up small corrie moraines there. Cwm Idwal in North Wales contains such a feature. In Scotland this readvance is called the *Perth Readvance* and was quite extensive, as the map shows. This period is generally known as Zone I, and is recognised throughout Europe.

A warmer phase followed (Zone II) and ice completely melted from the British mountains. Then about 12 000 years ago the climate cooled again giving small corrie glaciers south of the border, and fairly extensive valley glaciers, which constitute the *Loch Lomond Readvance* in Zone III in Scotland (Fig. 7.16).

This was the last hiccup of the ice age and the climate warmed steadily until about 5000 years ago when it was 1–2°C warmer than at present in Britain. This period is known as the *Climatic Optimum*. Cold phases have occurred since, for example the 'Little Ice Age' from 1750 to 1850, when the Thames froze hard enough to hold 'frost fairs' upon it. But it is doubtful whether even small glaciers built up on Britain's mountains at that time.

ASSIGNMENT

The sequence of deposits in Fig. 7.21 represents a record of the events during the Devensian glaciation at Trwyn y Tal on the coast of Caernarfon Bay about 20 km south-west of the town of Caernarfon. Interpret the section and build up a picture of the climatic events in this part of Snowdonia. The descriptions of the members of the sequence provide clues to help you do this.

Key Ideas

A. *Glacial erosion*
1. The problems of time scales, multiglacial episodes, the postglacial non-glacial processes since, and subglacial observations make it difficult to unravel fully the processes of erosion.
2. The effectiveness of glacial erosion depends much on the movement of the ice, the load the glacier carries and the hardness of the rocks it traverses.
3. *Abrasion* and *plucking* are the two most important processes of glacial erosion.

Fig. 7.21 The deposits at Trwyn y Tal, Caernarfon, Gwynedd (After Whittow)

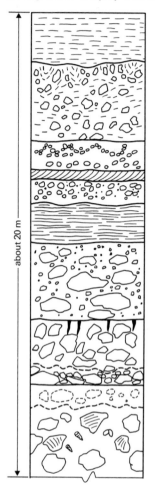

slope wash

local rock fragments and plant remains indicating climate slightly warmer than present

head

with cryoturbation (frost heave) structures in upper surface, occasional rounded erratic pebble in generally angular debris

fluvioglacial gravels

cryoturbation in upper layers, sandy layers with current bedding and ripple marks

laminated clays

varved silts and sands undisturbed by frost action

local till

contains no erratics, volcanics and granite-like rocks of Welsh origin, boulders up to 1m long with striations on surface

head

large blocks of local origin, some small frost wedges in upper surface

boulder pavement

few fine particles, some boulders striated

grey till

Calcareous, contains shelly debris of cold climate marine animals, erratics of Scottish and Lake District origin, few local boulders, weathered upper surface (decalcified)

about 20 m

4. *Pressure release* and *subglacial meltwater erosion* undoubtedly make some contribution as well.
5. *Glacial troughs* with *truncated spurs, hanging valleys, ribbon lakes* (often drained) with *cirques* and *arêtes* at their heads are characteristic landforms resulting from erosion.
6. Ice erosion affects both long and transverse profiles of valleys.
7. A *parabola* is a useful way to describe the shape of a glacial trough in conjunction with a *form ratio*.

B. *Glacial transport*
1. Debris is transported in three zones: *supraglacial*, *englacial* and *subglacial*.
2. Trails of *erratics* across the landscape are valuable evidence in determining the directions of ice flow.

C. *Glacial deposition*
1. Glacial drift is classified according to its manner of deposition: *till* (boulder-clay), *ice-contact stratified drift* or *outwash material*.

2. *Drumlin swarms* are useful indicators of the direction of ice flow.
3. Drumlins may radically alter postglacial drainage in relation to preglacial drainage.
4. *Elongation ratios* and *orientations* enable drumlins to be objectively measured and compared.
5. *Till fabric analysis — pebble orientation* and *roundness* — again tells us something of the glacial history of a locality.
6. *Ablation and lodgement tills, terminal* and *recessional moraines* and *lakeside erosional terraces* all contribute to the build-up of a full picture of the glacial episodes in a locality.

D. *Fluvio-glacial or meltwater deposition*
1. Meltwater deposition builds up an *outwash plain* with sorted, stratified and rounded deposits.
2. *Varve* sediments provide a useful method of calculating glacial time from their annual rhythmic sedimentation.
3. Outwash plains or valley trains are relatively steeply inclined as meltwater streams aggrade their channels to cope with the great volumes of sediment that are transported.
4. Ice-contact stratified drift features of *kames, kame-terraces* and *eskers* reveal something of the supraglacial, englacial and subglacial drainage patterns.

E. *Glacial meltwater effects*
1. Meltwater lakes overflowing to nearby valleys cut *spillways*.
2. There are several examples in Britain where *proglacial lakes* have modified preglacial drainage patterns.
3. Subglacial water flowing under hydrostatic pressure erodes *humped* or '*up and down*' channels.

F. *Britain during the Pleistocene glaciation*
1. Glaciers and ice sheets occurred in Britain during the *Pleistocene* period which consisted of a number of *glacial* and *interglacial periods*, reflecting climatic change.
2. Evidence is scattered, and nowhere does there appear to be a complete record of glacial fluctuations.
3. Evidence of earlier episodes has often been obliterated by later glaciations.
4. The record of the Pleistocene depends on evidence from a wide variety of sources — biological, geological and climatological, as well as geomorphological.
5. At any one time Britain has been occupied by a variety of environments through periglacial to boreal pine forest, reflecting the climatic changes associated with Pleistocene glacial and interglacial periods.

Additional Activities

1. With reference to the 1:25 000 map of the Glacier de Bonne Pierre (Fig. 6.16), and using the accompanying photographs (Plates 6.12 and 6.13):
   (a) draw labelled sketch maps of a horn, arêtes and a cirque.
   (b) Plate 6.12 is a lateral moraine (the footpath traces its course) and the

glacier snout lies to the right (south) almost completely covered by debris. Note the scale! Describe the lateral moraine in relation to its size, the steepness of its sides and the nature of the debris.

(c) From the crevasses pattern, conclude how the long profile of the valley floor may vary. Construct a long profile of the full length of the valley from the Col des Ecrins, 3367 m, to the confluence with the main valley at 1865 m. Use the spot heights on the glacier surface and glacier contours until you reach the snout. This should indicate the stepped profile that relates to crevasse formation.

(d) Construct a valley cross-profile immediately below the snout (at 2345 m). Describe its shape. Does this confirm your conclusion to additional activity 6(b) at the end of Chapter 6?

2. The data shown in Table 7.1 relates to Fig. 7.22, which shows the distribution of 56 cirques in Snowdonia. For convenience, they are grouped into four, according to their location, and are numbered with reference to the nearest highest peak.

(a) Construct a line graph with 'distance from south-western edge of map' in kilometres along the 'x' axis and 'elevation of cirque' in metres along the 'y' axis. Draw a line of 'best fit'. How do you explain the correlation? Alternatively, use a statistical method to test the correlation, e.g. Spearman Rank Correlation Coefficient.

Fig. 7.22 Cirque location within Snowdonia (Adapted from Joint Matriculation Board Practical Exam 1977)

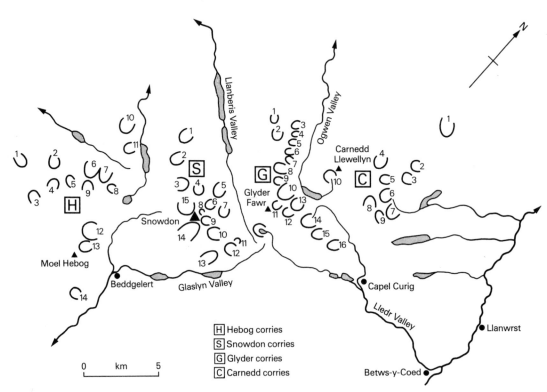

Table 7.1 Snowdonia cirque elevation (in metres) and orientation (in degrees from true north) (Adapted from Joint Matriculation Board Practical Exam 1977)

1. Hebog corries	Elevation	Orientation
H 1	270	310
H 2	385	330
H 3	285	120
H 4	400	140
H 5	385	70
H 6	335	340
H 7	335	345
H 8	370	85
H 9	420	155
H 10	300	345
H 11	400	20
H 12	285	40
H 13	385	10
H 14	335	95

2. Snowdon corries	Elevation	Orientation
S 1	320	10
S 2	335	30
S 3	470	240
S 4	635	310
S 5	420	20
S 6	835	40
S 7	750	0
S 8	935	80
S 9	670	85
S 10	470	70
S 11	420	175
S 12	370	80
S 13	350	190
S 14	385	180
S 15	570	320

3. Glyder corries	Elevation	Orientation
G 1	585	330
G 2	670	340
G 3	570	30
G 4	600	60
G 5	500	45
G 6	500	55
G 7	520	75
G 8	585	50
G 9	700	80
G 10	420	50
G 11	835	10
G 12	800	15
G 13	600	10
G 14	500	35
G 15	600	45
G 16	420	55
G 17	420	85

4. Carnedd corries	Elevation	Orientation
C 1	535	350
C 2	585	85
C 3	685	80
C 4	835	355
C 5	920	45
C 6	535	45
C 7	470	5
C 8	585	110
C 9	600	170
C 10	735	120

(b) Construct a rose diagram (see Fig. 7.8 for method) to illustrate the importance of aspect of the cirques, and comment on the result.

3. Use the Ordnance Survey 1:25 000 map (Fig. 7.23) to identify the shapes of drumlins. Use elongation ratios and orientation measurements to describe and compare them. The contour interval is small enough to define their shapes. What direction was ice flow? Suitable maps for similar studies in other areas would be of the Midland Valley of Scotland, the Eden valley and Solway Lowlands in Cumbria, the Lune valley in Cumbria or upper Ribblesdale in North Yorkshire.

4. Using the method suggested on page 192, try to establish a preferred direction of pebble orientation from a till fabric analysis. Measure between 30 and 50 pebbles. If no clear orientation occurs, suggest reasons why this may be so. Consider the following: if the sample is collected near the surface, mass movement either in periglacial or recent conditions may be the cause. It may also be the case that no 'streamlining' of the deposits occurred at all; they were simply deposited as the ice melted and their orientation is entirely at random.

5. Take measurements of between 30 and 50 samples of pebbles for roundness (method as indicated on page 117) from a glacial till. If possible, compare them with pebble roundness from other environments, e.g. scree slopes, solifluction material, fluvio-glacial deposits (Section D, page 196), river-borne deposits or beach deposits.

6. Suggest the differences in the depositional features that are formed under advancing and retreating ice sheets or glaciers.

7. Why is it difficult to be definitive in one's explanation of a landscape that has been glaciated?

Reading

BLOOM, A.L., *The surface of the earth*, Prentice Hall, pages 128–145, 1969
HANWELL, J.D. and NEWSON, M.D., *Techniques in physical geography*, Macmillan, pages 187–200, 205–208, 1973
DURY, G., *The face of the earth*, Penguin, Chapters 12, 13 and 14, 1970
STRAHLER, A.N., *Physical geography*, Wiley, pages 523–544, 1975
SAWYER, K.E., *Landscape studies*, Edward Arnold, pages 77–90, 1975
SPARKS, B., *Geomorphology*, Longman, Chapters 12 and 13, 1972
HOLMES, A., *Principles of physical geography*, Nelson, pages 402–444, 1978
PITTY, A.F., *Introduction to geomorphology*, Methuen, pages, 248–255, 372–380, 1971
SMALL, R.J., *The study of landforms*, Cambridge University Press, pages 361–416, 1978
MONKHOUSE, F.J., *Principles of physical geography*, University of London Press, Chapter 8, 1975
For advanced studies the following provide excellent summaries of much of the current research:
*PRICE, R.J., *Glacial and fluvioglacial landforms*, Longman, 1976
*SUGDEN, D.E. and JOHN, B.S., *Glaciers and landscape*, Edward Arnold, 1976
*EMBLETON, C. and KING, C.A.M., *Glacial and periglacial geomorphology*, Edward Arnold, 1975
*JOHN, B.S., *Winters of the world*, David and Charles, 1979

Fig. 7.23 (*opposite*) Ribblesdale, West Yorkshire. Extract from Ordnance Survey 1:25 000 map, Sheet SD 95 (First Series), reproduced with permission of the Controller of Her Majesty's Stationery Office, Crown Copyright Reserved.

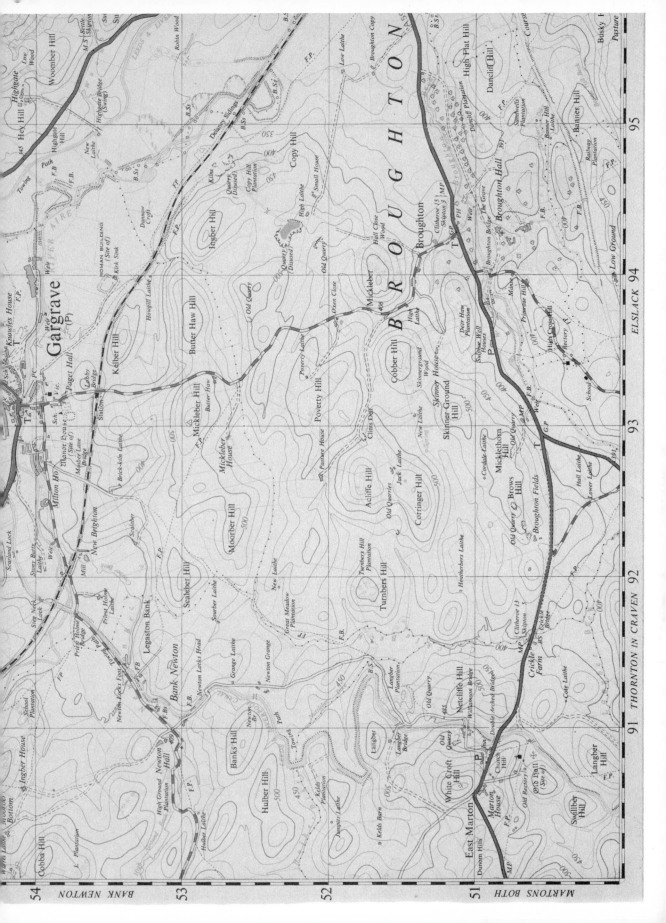

8 Periglacial environments

The word periglacial literally means 'fringe of the glacier', but it is used to indicate all environments which develop in cold climates, except glaciers and their related features. The frost-shattered peaks of the Alps, the almost permanently frozen tundra areas of Arctic Russia and Canada and the tops of the higher mountains in Britain are all areas where different types of periglacial processes are found today. During the glacial period the extent of the periglacial zone was much greater and in the temperate areas of the world fossil periglacial landforms are just as widespread as fossil glacial landforms. Periglacial landforms, in short, can be found wherever freezing occurs often enough and hard enough to push other processes into second place.

A. The Permafrost

There are large areas mainly north of the Arctic Circle where the temperature a few metres below the surface never rises above 0 °C. This perpetually frozen layer is known as the *permafrost*. It may be regolith or bedrock. It may be 'dry' with no ice or it may contain large amounts of ice. At present, approximately one fifth of the world's land area is underlain by permafrost. Most of this is in the northern hemisphere (Fig. 8.1a) but there are small areas in South America and ice-free parts of Antarctica where it can also be found.

The thickness of the permafrost layer varies. The cross-section in Fig. 8.1b shows how it thins out beneath the sea and towards the south where it becomes broken. The line of this cross-section is marked on Fig. 8.1a. Where temperatures are very low in winter, down to –50 °C, and the summer very short, so that the annual average temperature is less than –5 °C, the permafrost is *continuous*. In places it is up to 600 m thick. The permafrost is *discontinuous* where the annual average temperature is approximately between –5 and –1.5 °C. Here it is thinner and more broken. *Sporadic permafrost* is found where temperatures average between –1.5 and 0°C. In these zones frozen ground is only found in small isolated patches where the local climate is relatively cold for the latitude. The distribution of these three zones of the permafrost is shown in Fig. 8.1a. Notice how far south it extends in mountain regions where temperatures are lowered by altitude and in the centre of the large Eurasian landmass where very low winter temperatures down to –50 °C regularly occur. In the sporadic zone the permafrost is absent beneath the rivers.

Fig. 8.1 Permafrost: a) distribution and   b) cross-section along 130°E (After West)

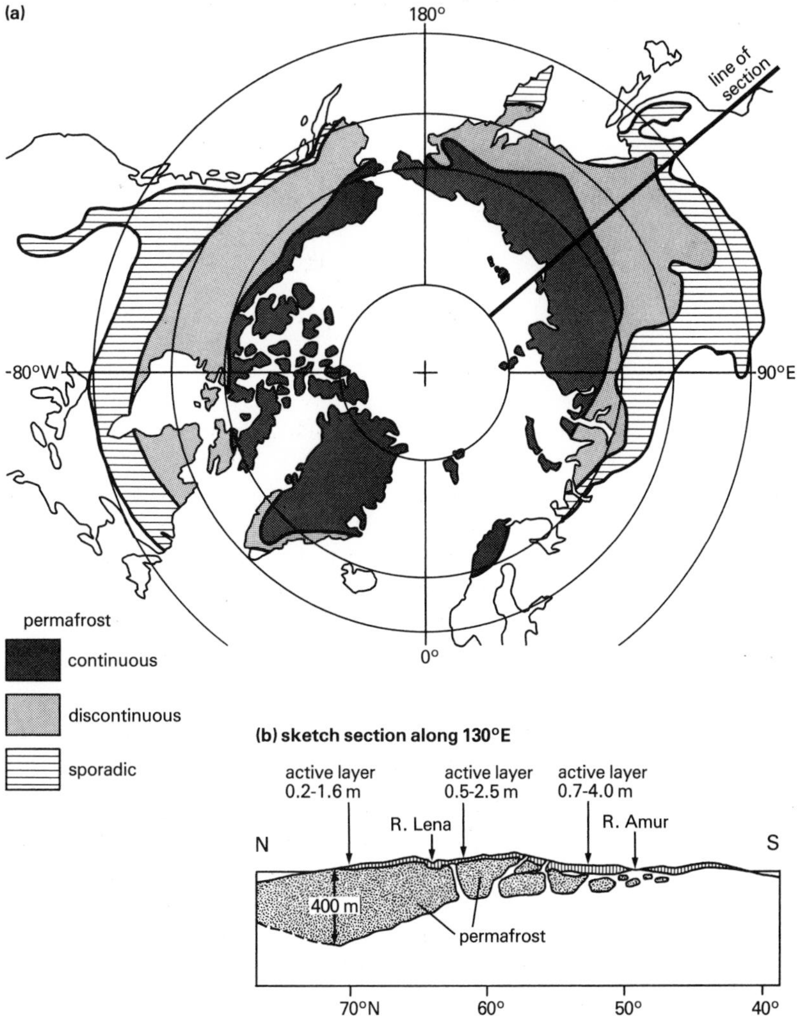

(a)

permafrost

■ continuous

▨ discontinuous

▤ sporadic

(b) sketch section along 130°E

active layer
0.2–1.6 m

active layer
0.5–2.5 m

active layer
0.7–4.0 m

R. Lena

R. Amur

N   S

400 m

permafrost

70°N   60°   50°   40°

1. Landforms resulting from movement of the active layer

The *active layer* is the name given to the layer above the permafrost which in summer has a temperature above freezing. The long hours of daylight at high latitudes counteract the limiting effect of low sun angles, and daytime temperatures can rise to quite high levels during the summer months.

The penetration of this heat into the ground is slow and although surface temperatures may rise above 0 °C in late April, at 2.5 m down the ground may remain frozen, except for a brief period in September after a whole summer's heat input. The depth to which the ground melts is dependent on the amount of heat received, which is largely a function of latitude and the thermal properties of the ground. Saturated ground may take far longer to rise above 0 °C than dry ground. In Fig. 8.1b the depth of the active layer varies from 0.2 m in the north to 4 m in the south. The actual depth depends on local conditions.

The melting of the ice in the active layer releases large volumes of water which is unable to drain away in the flat areas because of the frozen permafrost beneath. The saturation of the active layer gives rise to a variety of landforms.

On slopes as low as 2° the wet active layer can start to move in the same way as a mudflow (page 57). Under periglacial conditions such movements are given the general name of *solifluction* and can produce solifluction terraces. These are step-like features with 'risers' up to 15 m high and 'treads' up to 500 m long. The solifluction sometimes occurs under a vegetation mat which is pushed forwards and rolled under like a caterpillar track on a bulldozer (Fig. 8.2a). This produces a *turf banked lobe*.

Fig. 8.2 Turf banked solifluction lobes (a) and stone banked solifluction lobes   (b)

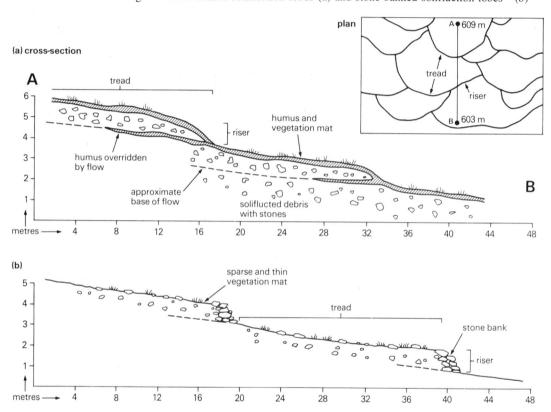

Stone banked terraces form where the vegetation cover is less continuous and stones more frequent in the active layer. The terraces tend to be smaller, up to 5 m high and 50 m long (Fig. 8.2b). The stones accumulate along the lobate front of the terrace because winter freezing pushes them to the surface (page 222) and moves them down the slope when the terrace in general is frozen solid. Although both these features are common in permafrost zones, they form in other areas where winter temperatures are low. They can be found at about 800 m in Scotland on Liathach and in the Cairngorms.

Fossil solifluction deposits are widespread in Britain. They are known by a number of names: *head* in areas of Paleozoic rock in south-west England, and

216

coombe rock in chalk areas of south-east England. Plate 8.1 shows a head deposit in Devon which has been disturbed by frost action (page 221). Where the particles are elongated, the long axes become orientated with the flow which is usually downslope. This is often useful to distinguish periglacial from glacial deposits which usually show orientation along the major valley feature (page 192). Many fossil solifluction terraces have been recognised in Britain. Plate 8.2 shows one in South Wales.

Plate 8.1 Cryoturbation in head, north Devon (Author's photograph)

2. Landforms resulting from freezing

(a) *Patterned ground*

(i) *Ice wedge polygons*

Air photographs of tundra areas frequently show network patterns on the surface (Plate 8.3). The largest nets, up to 50 m across, are only found where the

Plate 8.2 Solifluction terrace, head of Towy Valley, South Wales (Author's photograph)

Plate 8.3 Polygonal ice wedge pattern, Old Crow Plain, Yukon Territory (Geological Survey of Canada, Ottawa)

temperature in winter falls below –20 °C and where the permafrost is continuous. These nets are formed by ice wedges which are almost vertical sheets of ice, usually 1–2 m across at the top, tapering downwards (Fig. 8.3a). Some extend down 10 m before becoming too thin to be recognisable.

The key to their formation lies in the intense winter cold which in frozen silts and peaty tundra 'muck' sediments produces contraction cracks (Fig. 8.3b). Dry, muddy lake beds also develop cracks in this way which form a polygonal pattern on the surface. When surface temperatures rise above 0 °C water flows into these open cracks and freezes when it reaches the frozen ground beneath. This permafrost layer retains the ice until the following winter when the active layer and permafrost again contract giving a crack which is filled with water during the thaw. The annual layers of ice are generally less than 10 mm across, so that an ice wedge 1 m across would take at least 100 years to form.

Fig. 8.3 Ice wedge formation (a) and landforms developed in zones of ice wedges (b)

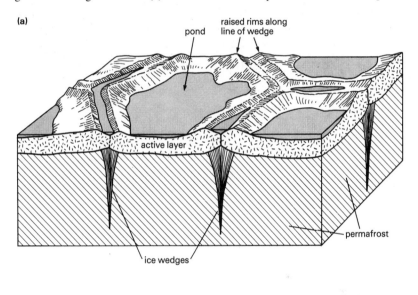

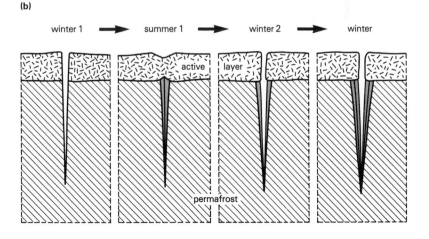

The patterns seen on air photographs are depressions in the active layer above the wedge. They are frequently water filled (Fig. 8.3a). Often these depressions have a raised edge and the centre of the polygons are covered by water. These raised rims are produced by expansion in the summer pushing up the sediments in contact with the ice wedge.

When the ice wedge melts, the area occupied by the ice may become filled with sediment. This is then called an *ice wedge cast* (Plate 8.4). In areas of Britain mainly outside the area affected by the last glaciation, fossil ice wedges can be detected from the air in *crop ripening patterns* (Plate 8.6). Small differences in the textures of the soils within the polygon and along the line of the wedge cause crops to ripen at different times. The wedge shows up as a green (dark) area, while the centre of the polygon where the crop is ripe is golden (light) in colour.

Plate 8.4 Ice wedge cast in stratified sands (Author's photograph)

220

Plate 8.5 Sorted stone polygons, Jotenheimen, Norway (Author's photograph)

(ii) Sorted stone polygons

Much smaller patterned ground features consisting of circles or *polygons* usually less than 10 m across are found today actively forming in permafrost and mountain areas (Plate 8.5). The edge of the polygon is formed of stones (pebble and boulder size), while the centre contains much finer debris (fine sand and silts) (Fig. 8.4).

The sorting of material can be explained in many ways. As expansion of the sediment occurs upwards on freezing, the stone is also lifted by its upper surface which is frozen to the expanding silts (Fig. 8.5). A large stone in a silty deposit responds more rapidly to temperature changes than the sediment around it, so that when freezing takes place from the surface downwards ice forms

Fig. 8.4 Sorted stone polygons and stripes

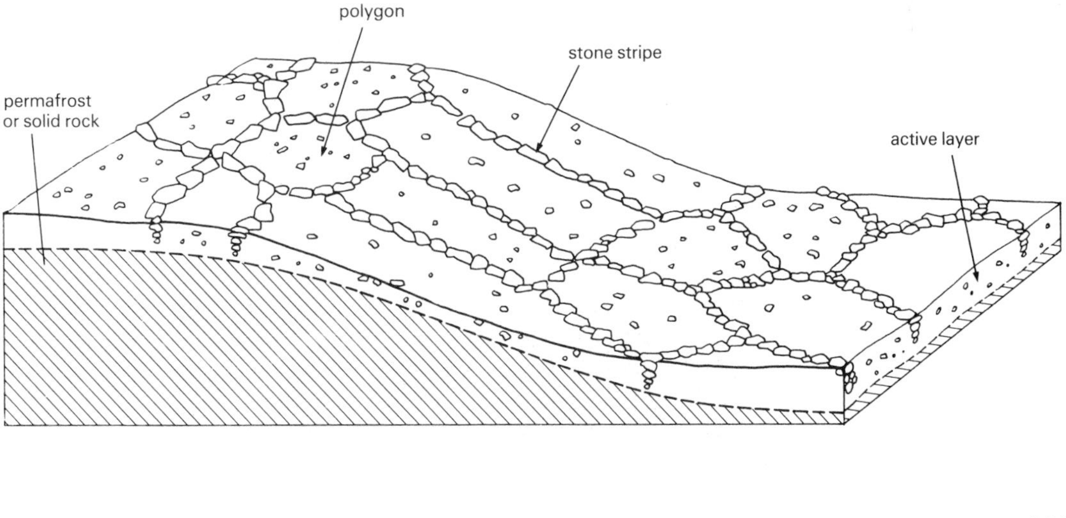

polygon

stone stripe

permafrost or solid rock

active layer

Fig. 8.5 The mechanism of frost heave (After Beskow)

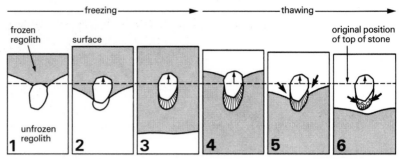

1 Freezing progresses down from surface accompanied by upward expansion, lifts stone by first freezing onto upper surface

2 Freezing front progresses down, space created below stone as it is lifted

3 Ice crystals grow into space and push up stone

4 Thawing progresses down from surface, contraction and lowering of surface level but stone still supported by ice crystals

5 Thawed sediment collapses around stone supporting it

6 Stone held in thawed sediment while ice crystals melt and space filled by collapsing sediment

most rapidly beneath the stone and pushes the stone up. When thawing begins it starts at the surface and the silt collapses round the stone which is still frozen in position at its base and prevents it falling back to its original position. In a hard winter in Britain it is possible to see the effects of this as stones in garden soils are pushed up to the surface.

Freezing produces the greatest amount of expansion in the finer deposits at the centre of the circle, so that stones on the surface are pushed progressively outwards to form the edge of the stone circle. Where a number of circles are forming close together they interfere with one another and form polygons.

In flat areas these polygons are nearly symmetrical but on a slope they begin to become elongated because the stones tend to move down the slope. Eventually the polygon is so elongated that it becomes a *sorted stone stripe* (Fig. 8.4).

Polygons and stripes can be seen in crop ripening patterns particularly in East Anglia (Plate 8.6) and on high mountains in Britain such as Helvellyn and in the Cairngorms. In the Tinto Hills, Lanark, sorted features which were dug up reformed in one year illustrating that the present climate is sufficiently cold for their formation.

(b) Pingos

In the Mackenzie Delta, Northwest Territories, a zone of continuous permafrost, there are a large number of almost circular hills. These are up to 300 m across and 60 m high and some have a central crater which may contain a circular pond. They are called *pingos*, which is the eskimo name for hill. They tend to occur in groups in almost flat areas. At the centre of each pingo is a core of ice (Plate 8.7).

222

Plate 8.6 Stone stripes and polygons seen in crop ripening patterns near Thetford, Norfolk (Cambridge University Collection — copyright reserved)

Plate 8.7 A pingo in Northwest Territories (Geological Survey of Canada, Ottawa)

This core is *segregation ice*. Water is drawn or segregated from the sediment and migrates towards a freezing centre where ice develops. It builds up at the centre as a large ice lens pushing up a dome of sediment above it (Fig. 8.6). The dome grows and eventually the cover of sediment splits and the ice core becomes exposed and subject to melting. In the Mackenzie Delta most pingos form in old lake beds and have a plentiful supply of water. In other areas they occur at the foot of slopes which supply water as the surface melts.

Fig. 8.6 Pingo formation

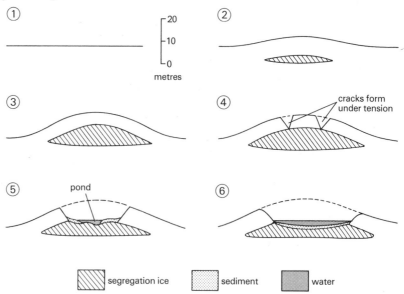

In Britain the only evidence of pingos are circular ramparts with infilled ponds. In Norfolk they probably formed at the coldest phase of the last glaciation (page 206).

ASSIGNMENTS

1. *Explain why pingos and ice wedges are found only in areas of continuous permafrost, while sorted stone features and solifluction terraces can occur in other areas.*
2. *Describe how you could use the methods for assessing particle shape and orientation to distinguish between periglacial (solifluction) and glacial deposits (page 192).*
3. *Discuss the factors which would favour the formation of turf banked lobes rather than stone banked lobes.*

B. Frost Weathering Features

Frost plays an important part in weathering bedrock both within permafrost areas and outside it. The effect of frost was described in Chapter 2; it produces scree and scree cones in mountain areas. In flatter areas bedrock becomes dis-

224

Plate 8.8 Blockfield, Glyders, North Wales (Author's photograph)

rupted and the shattered debris accumulates to form a *blockfield* (Plate 8.8). Extensive blockfields are found on many mountains in Britain. Glyder Fawr in North Wales has a very large blockfield at Drws-y-Gwnt (Castle of the Winds).

C. Features Formed by Snowbanks

1. Nivation

At the beginning of this chapter we mentioned that periglacial areas have a discontinuous snow cover. However, perennial snow banks do occur in favourable locations and beneath them *nivation* occurs (page 183). Frequently this process does not lead to the development of a fully-fledged glacier and the hollow produced by nivation alone is called a *nivation cirque*. Cwm Du shown in Fig. 8.7a is a fossil nivation cirque. The majority of these features in Britain are smaller than this and it could be argued that Cwm Du is too large merely to have been a snow patch feature. In contrast to true cirques, the floor always slopes outwards and in general nivation cirques are much smaller. Many small hollows on mountains in Britain were probably formed by nivation.

2. Protalus

On steep slopes debris from frost shattering may fall onto a snow patch and slide to its foot to form a ridge (Fig. 8.7b). These are called *protalus ramparts*

Fig. 8.7 Nivation cirque (a) and protalus ramparts (b) (After Watson)

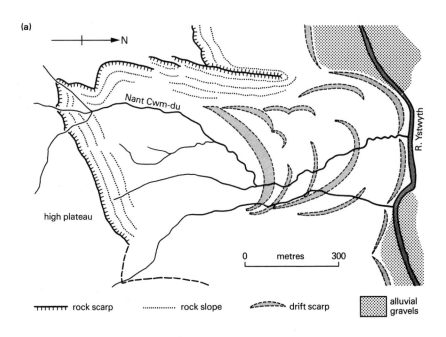

(a)

N

Nant Cwm-du

high plateau

R. Ystwyth

0   metres   300

πππππ rock scarp   ··············· rock slope   ◠◠◠◠◠◠ drift scarp   ▦ alluvial gravels

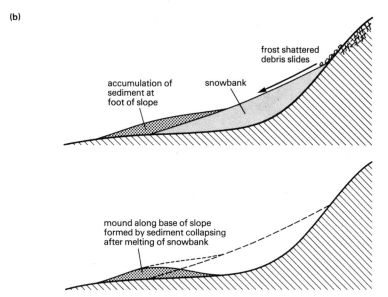

(b)

frost shattered debris slides

accumulation of sediment at foot of slope

snowbank

mound along base of slope formed by sediment collapsing after melting of snowbank

and one can be seen on the north side of Cwm Idwal in Snowdonia. They are probably quite common as late features in many corries.

D. Periglacial Wind Deposits

In dry environments the wind plays a significant part in producing landforms. The fine sediment from glacial and periglacial deposits which have dried out may be removed by the wind and deposited as dunes. During the *Devensian* layers of wind-blown material, called *loess*, up to 15 m thick were deposited in northern Europe and northern America. Loess is frequently calcareous and forms the basis of a productive agriculture. In East Anglia the brick-earth deposit is a loess.

Loess forms a blanket deposit which tends to mask the relief of the area. In Holland, to the south of the ice front of the last glaciation, huge *sandur* or outwash plains were subject to wind action. Dunes formed which were parabolic or 'U' shaped with their convex side pointing down wind (Fig. 8.8).

Fig. 8.8 Sand dunes in the coversands of Holland

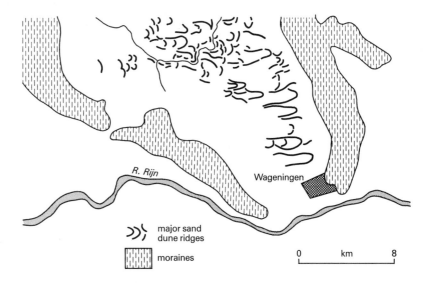

These *coversand* deposits, which are the coarse grained equivalent of loess, indicate strong winds towards the east. Much of the finer silt and clay material was carried away towards the east and deposited as loess in Eastern Europe. Conditions must have been very similar at Chelford during the Devensian (page 206). In southern Iceland today dust storms sweep across the sandur plains winnowing silt and clay from the outwash of the Vatnajökul ice cap.

ASSIGNMENTS

1. *With the aid of labelled diagrams, explain the features you would look for to distinguish between a protalus, a nivation cirque and a glacial cirque.*
2. *Explain why it is thought that the periods of loess deposition in Europe must have been dry as well as cold. Refer to page 107 and river erosion before you answer this question.*

Key Ideas

A. *The permafrost*
1. Permanently frozen ground, *the permafrost*, may be continuous, discon-

tinuous or sporadic. This distribution is dependent both on present and past climate.

2. The *active layer* is unfrozen during the summer and forms a highly mobile layer liable to movement on very gentle slopes as a result of its low angle of internal friction.

3. Very cold climates with extensive winter freezing develop *ice wedges* and *pingos*. Freezing also sorts deposits by grain size.

B. *Frost weathering features*
1. Frost shattering is very effective in climates which fluctuate frequently around freezing point. This effect is not confined to periglacial areas.

C. *Features formed by snowbanks*
1. Snowbanks develop distinctive landforms. Beneath them *nivation* can destroy the bedrock.
2. Above snowbanks sliding debris forms *protalus*.

D. *Periglacial wind deposits*
1. *Loess* is the product of wind deposition mainly during the Pleistocene. Its sandy equivalent in Europe is called *coversand* which forms dunes.

Additional Activity

Place layers of sand, silt and gravel in a plastic container. Moisten, but do not saturate, with water and place in a freezer. Allow it to freeze completely and then remove it and let it thaw. Repeat this about 20 times and then cut the sample across vertically and compare the layering with the original pattern. Silty sediments give the best results. It is possible to use pebbles to study frost heave in the same way.

Reading

WEST, R.G., *Pleistocene geology and biology*, Longman, pages 71–107, 1977
*RICE, R.J., *Fundamentals of geomorphology*, Longman, pages 293–311, 1977
*EMBLETON, C. and KING, C.A.M., *Glacial geomorphology*, Edward Arnold, 1975

9 Coasts

A. Introduction

The coast is a narrow overlap zone between the land and the sea in which the processes of each operate. In preceding chapters we have examined the role of rivers, slopes and glaciers. Each of these, as Fig. 9.1 indicates, contributes sediment to the coast. In general the sea contributes little material but an enormous

Fig. 9.1 Energy and sediment sources for the coastal system

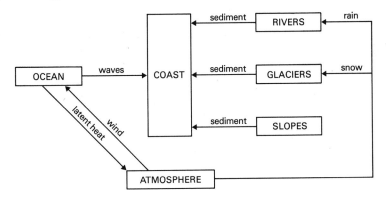

amount of energy to the coast in the form of *waves*. Waves are the most noticeable features of the sea's surface. They are the outcome of the action of the wind. Less obvious are the *tides* which cause the surface to oscillate once or twice a day. These are the result of the gravitational attraction of the sun and the moon.

The waves and tides are used to divide the coast into zones (Fig. 9.2). The

Fig. 9.2 The zonation of the coast

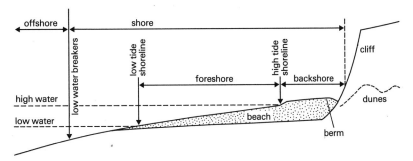

landward limit of the *shore* is the limit of wave action under exceptional conditions of storm or high tides. Seaward of the *low water breaker line* lies the *offshore* zone. The *backshore* lies above high water and the *foreshore* between the two tidelines.

1. Waves

The oceans cover about 70 per cent of the earth's surface, and, apart from those areas which have an ice cover, much of this water is agitated into waves for most of the time. The wind blowing over the smooth surface of the water instigates very small ripples which grow into recognisable waves. If the wind continues to blow for several hours the *wave height*, the difference between the trough and crest (Fig. 9.3), increases. A wind of 50 km/h blowing for 30 hours

Fig. 9.3 Basic terms used in the description of waves

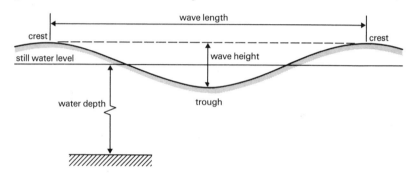

over the open sea would give waves about 6 m high. In coastal areas the sea is not entirely open. For example, the coast of Norfolk is separated from Holland by only 200 km of open water, which is an insufficient distance for an east wind to develop waves to their full height. However, there are thousands of kilometres of open water stretching north-east towards the Arctic Ocean and much larger waves approach from this direction. The length of open sea over which the wind blows to generate waves is the *fetch*. On a pond or lake the effect of fetch is very easy to see. The downwind end will have very small waves or be almost calm in comparison to the upwind end which will have larger waves. In Cornwall the longest fetches are to the south-west and the largest waves approach from this direction.

(a) *Waves in deep water*

You have probably seen quite large waves breaking on the beach when there has been little or no wind. These waves are called *swell* and have often travelled great distances from the areas where they were generated. Waves reaching the coast of Cornwall have been attributed to storms in the South Atlantic in the region of Cape Horn, about 13 000 km away. Waves in the area where the generating wind is blowing are called *sea waves*, or simply *sea*. They are much less regular in size and direction than the swell waves and reflect the nature of the wind which generates them. The waves fan out to 20° on either side of the wind direction which produces them. The wind creates many different wave types: some are short, only a few metres from crest to crest; others are long. The long waves travel much more efficiently through the water and overtake the shorter ones, so it is this type of wave that is recorded at long distance.

If you have been in a small boat in a swell you will no doubt have noticed the waves advancing towards you; they lift the boat up and let it down while moving it very little over the sea surface. Objects floating near the boat move neither significantly towards nor away from it as the wave passes. The water itself does not move. It is only the shape of the wave that passes through the water. Board surfers take advantage of this phenomenon. They slide down the sloping front of the wave and plane like speedboats over the almost stationary water. If the nose of the board dips under the water surface, it stops dead in the still water and the surfer is catapulted forwards, clear evidence that the water is static!

The shape of waves in deep water is called *trochoid*, which is the shape described by a fixed point on a circle as the edge of the circle is rolled along a straight line (Fig. 9.4). If we imagine the circle centre to be fixed and the wave

Fig. 9.4 The motion of a particle as a wave passes

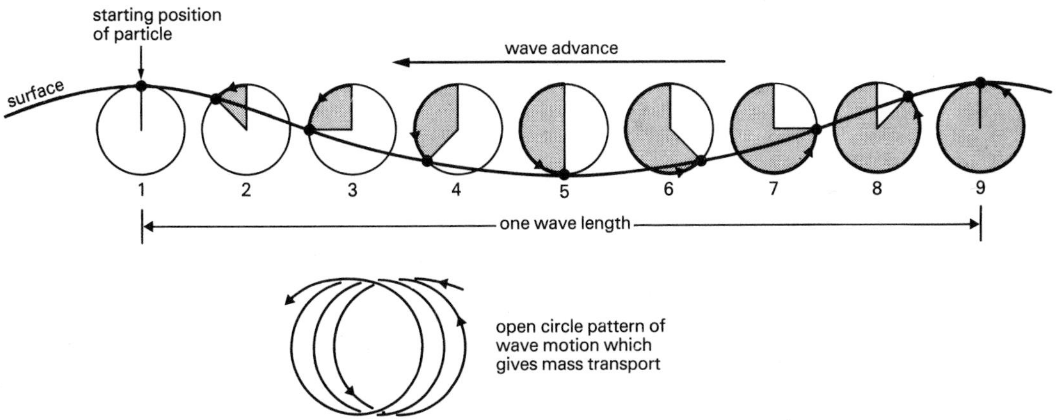

shape to be moving, then the point must describe a circular orbit as the wave shape passes. It first moves to position 1 in Fig. 9.4 and then forwards and down to position 5 in the trough. A backwards and upwards movement occurs as the next crest advances and the particle is lifted back to its original position, thus completing its circular orbit. This perfect wave pattern is rarely observed in nature and usually there is some forward transport of particles of water. This is called *mass transport* and it is very small in relation to wave velocities.

The time taken for successive crests to pass a point is called the *wave period* and is almost constant despite other changes in the wave. The *length* of a wave (L) is equal to the product of the *period* (T) and the *velocity* of the wave (C):

$$L = CT$$

We have already said that long waves travel faster than short ones. Consequently they have longer periods and this is how they are distinguished from other shorter waves on reaching the coast. Long waves which have travelled hundreds of kilometres may have periods of up to 20 seconds. Local waves have periods of 5 to 8 seconds. A group of waves is called a *train*. Trains of waves reaching the coast can be observed from high viewpoints such as clifftops or aircraft.

The energy of a wave is made up of two parts. Half is *potential* energy resulting from the position above the trough, and half is *kinetic* caused by the motion of the water particles within the wave. Both these types of energy have been mentioned before (page 92). In total, the energy of the wave train travels at half the speed of the wave form, and waves, if watched closely in favourable conditions, will be seen to rise at the rear of the train and pass through to the front of the train where they die out.

Long waves with a low height may be barely perceptible at sea as the gradient or *steepness* of a wave is very low. Short, high waves are steep and therefore much more noticeable even though they transport far less energy. The difference between the two becomes more apparent when the wave enters shallow water. The long, low swells of the open ocean become the *surf waves* characteristic of exposed coasts with long fetches.

(b) *Waves in shallow water*

The sea is considered deep when the water depth (Fig. 9.3) is greater than the wavelength of the wave. The wave motion does not extend to this depth. The trajectories of particles beneath the surface in deep water are circular or nearly so, but the diameter of these circles becomes smaller as the distance from the surface increases (Fig. 9.5). For each ninth of the wavelength the diameter of the circle is halved, therefore at a depth of one wavelength the circle is 1/512 of its surface diameter. A swell wave 100 m long and 1 m high in 100 m of water would produce a particle orbit of less than 2 mm and would be unlikely to affect the sea bed to any great extent.

Fig. 9.5 The movement of particles in a wave: a) in deep water and   b) in shallow water

(a) deep water   (b) shallow water

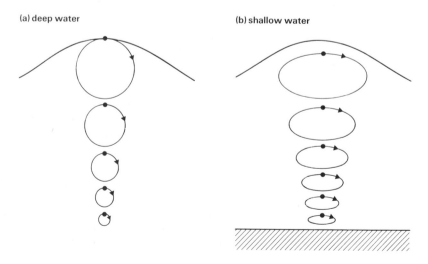

The velocity of a wave is not significantly affected by the depth of water until the depth is half the wavelength. As the depth falls below half the wavelength, friction with the sea bed begins to slow down the wave and change its characteristics. When the water is very shallow, the velocity is determined solely by the water depth. This means that all waves in shallow water travel at about the same speed. Friction with the sea bed absorbs some of the energy transported by the wave, but as the wave is slowed down energy is diverted into other changes in the wave. Figure 9.6 shows how the height and the wavelength adjust to shallow water. The first real changes begin to occur when the depth is half the wavelength. The height diminishes to about 0.90 of its deep water value, and when the depth is 0.06 of the original wavelength the height begins to increase rapidly. If you watch swell waves they appear to 'climb' out of the water as they approach the beach. They also appear to catch one another up, which is shown on Fig. 9.6 by the rapid decrease in wavelength in shallow water.

Fig. 9.6 Changes in the dimensions of waves entering shallow water (After King)

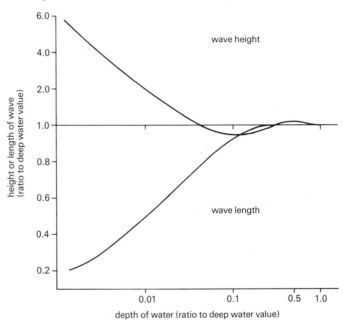

If the height increases and the length decreases, then the wave becomes steeper, since steepness is the ratio of height to length. The limit to the steepness of a wave is about 1 in 7 giving a crest angle of 120°; beyond this point the wave breaks. At the break point the deep water, orbital motion of the particles has become compressed to a very flat elipse (Fig. 9.5) in which the landward-moving upper part is accelerating and the seaward-moving lower part is decelerating. Eventually the velocity towards the land exceeds the velocity of the wave form and the water rushes beyond the wave, translating the potential energy into kinetic as the wave breaks. This energy change is the single most important event in the coastal zone from the geomorphologist's point of view. This is how waves do *work* on beaches and against cliffs.

(c) *The breaking wave*

There are various ways in which a wave breaks. One is the *plunging breaker* in which the crest moves forwards and downwards to enclose an air space for an instant between the wave and the broken water (Fig. 9.7). These occur most

Fig. 9.7 Plunging and spilling breakers

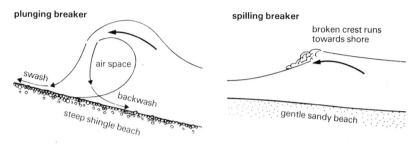

frequently when slow waves approach a steep beach. The swash or upbeach component is less important than the downbeach component, the *backwash*. At the other end of the series is the *spilling breaker* in which most of the breaking water is directed upbeach. The swash and backwash are primarily responsible for the movement of beach material but we will see that there are other factors involved (page 249).

2. Tides

The tides are the oscillations of the sea surface which occur regularly, with a period of about one day. They are caused by the gravitational attraction of the moon and the sun on the earth's water surface. The sun is so far away, 93 million miles, that despite its enormous mass it exerts less gravitational force on the earth than the moon, which is about a quarter of a million miles away. The moon is responsible for six-elevenths of our tides.

Great Britain has approximately two high tides a day, 12 hours 25 minutes apart. They are often of unequal height. *Spring tides* are exceptionally high or low tides. They occur shortly after each new and full moon and so are about 14 days apart. *Neap tides* occur midway between spring tides, just after each first and third quarter. They have the smallest range during the cycle.

The tide behaves like a wave which oscillates round a shallow tray of water when you try to carry it. These waves rotate about points at which there is little or no tidal range, called *amphidromic points*. The *range* of the tide increases radially outwards from these centres. Figure 9.8 shows the location of these points for the North Sea. They are situated nearer the coasts of Holland and Norway than of Britain, consequently those coasts have much smaller tidal ranges. The *cotidal* lines which radiate from the amphidromic points indicate the difference in the time of high tide in hours. This information for Britain is also shown in the AA Handbook. High tide at Hartlepool occurs 5 minutes after Sunderland which lies to the north. Whitby's high tide occurs 14 minutes later and Scarborough's 34 minutes after that. The tide is like a wave sweeping southwards along the coast.

Fig. 9.8 Cotidal and range lines in the North Sea (After King)

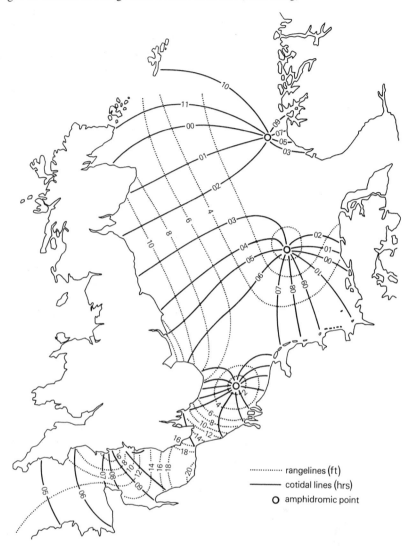

rangelines (ft)
cotidal lines (hrs)
O amphidromic point

In some parts of the world large tidal ranges are accentuated by the shape of the coast. The Severn Estuary and the Bay of Fundy in Nova Scotia, because of their funnel like shapes, constrict the tide and cause the amplitude to increase as the tide sweeps into the narrower part of the embayment. The advancing tide develops a steep front and in the case of the Severn causes a wave moving at about 30 km/h to travel almost to Gloucester. Such waves, or *tidal bores*, are not uncommon. They occur on the Rance Estuary in Brittany, in the mouth of the Colorado and in Morecambe Bay.

3. Wave and tidal environments

The winds which are responsible for the generation of the largest waves are located in two latitudinal belts which correspond to the position of the Polar

Front zone in each hemisphere. In these zones, located at 45 to 55° N and around 55° S, winds of high velocity are common around depression systems. The zone in the northern hemisphere fluctuates from north to south with the seasons; depressions tend to be more frequent in the winter, hence storm wave activity is at its peak in that season. The southern hemisphere belt is more stable in location as a result of the relatively small area of land in the south. Waves generated in these areas travel out as swell to affect coasts possibly thousands of kilometres away (page 230). Thus there are storm wave and swell wave environments. There are other zones where wave generation occurs (Fig. 9.9). The tropical easterlies or trade winds are on average less strong than those

Fig. 9.9 Major world wave environments (After Davies)

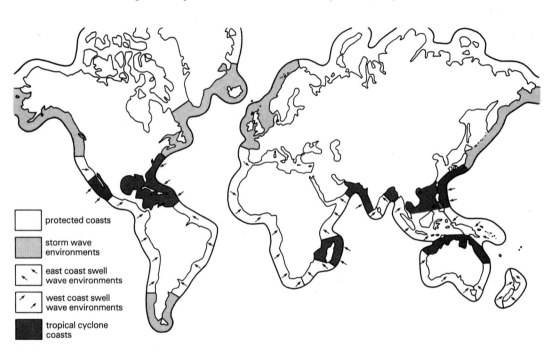

protected coasts

storm wave environments

east coast swell wave environments

west coast swell wave environments

tropical cyclone coasts

of the Polar Front zone, but they still generate waves which have a significant geomorphic effect. The monsoon is really a strengthened trade wind and in parts of Asia this is very significant at particular times of the year. Figure 9.9 shows the world pattern of wave environments in a generalised way.

We have seen that tidal ranges vary from place to place (page 234). This is important in that it spreads the effect of the waves over different ranges, and different landforms result. The pattern of tidal ranges is shown in Fig. 9.10. Those coasts with ranges greater than 4 m are called *macrotidal*, those with a range between 2 m and 4 m, *mesotidal*; smaller ranges are *microtidal*. It is interesting to note that microtidal environments are common on open ocean coasts. The tidal range in an embayment is often large for the reasons explained on page 235.

236

Fig. 9.10 World tidal environments (After Davies)

macrotidal >4 m

mesotidal 4-2 m

microtidal <2 m

ASSIGNMENTS

1. *The formula on page 231 relates length, period and velocity. Describe in words what happens (a) as the length decreases, and (b) as the period increases.*

   If a wave has a period of 10 seconds and a length of 50 m what is its velocity?

2. *What are the maximum fetch lengths and directions of the coast at: i. Blackpool, ii. Spurn Head, iii. Lands End, iv. Dover?*

   On a map of Britain mark areas where the fetch from any direction may exceed 1000 km and where the maximum fetch is less than 100 km.

3. *Using Fig. 9.6, describe the changes which would take place in a wave with a deep water wavelength of 60 m and a height of 1 m, when the water depth was half, quarter, one tenth and one twentieth of the wavelength.*

4. *Construct a table to show the 12 possible combinations of wave and tidal environments.*

   Using Figs. 9.9 and 9.10, mark on this table the type of coastal environment found in Patagonia, the north coast of Alaska, south-west Australia and the east coast of Ethiopia.

B. Coastal Erosion

1. Erosion by waves

There are many similarities between the way in which rivers erode (page 105) and waves erode, in that they both depend on the movement of water. However, the velocity required to achieve this erosion is generated in different ways. We have seen that there is little movement of water in waves in the open sea and it is not until the waves break on the shore that the water moves rapidly enough to erode (page 233). The rapid movement of water towards the shore has three main erosional effects: wave quarrying, abrasion and attrition.

(a) *Wave quarrying* affects previously loosened or unconsolidated rock fragments. A wave breaking against a cliff exerts a considerable impact or *shock pressure* as perhaps hundreds of tonnes of water hit the rock face. The energy of the wave is proportional to its height (page 232). Large storm waves on the French coast have been recorded with pressures up to 50 kg/cm^2 (pressures in car tyres are about 2 kg/cm^2). This high water pressure exists for only an instant but it may be sufficient to loosen blocks by acting along fault, joint and bedding planes. It is analogous to hydraulic action in river erosion. A plunging breaker may trap air against the cliff and a pneumatic pressure may assist in loosening blocks. Such an effect can extend the erosive power above the water line.

(b) *Wave abrasion* is the most effective facet of wave erosion. The wave, armed with pebbles, sand and, in the largest storms, boulders, is hurled against the cliff. The effect is to wear away the rock in the zone where this movement occurs. In hard rocks the action is slow and produces smoothed areas. In alternating hard and soft rock types small differences in abrasion resistance become accentuated over time with the harder strata becoming more prominent and so more subject to abrasion. In soft rocks abrasion may be so effective that the eroded material forms a protective mantle over the solid rock.

(c) *Wave attrition* affects the detached materials produced by quarrying. As in rivers, the particles become rounded and reduced in size as they collide with one another (page 105). Bricks, which are relatively soft, are a good indication of this and it can be useful to compare wave attrition by watching how fast rounding takes place. Most of this rounding takes place in the breaker zone. The movement of the tide has the effect of extending the range of the attrition vertically, so that at high tide the particles at the high tide shoreline are affected and 6 h 12 min later it becomes the turn of those at the low tide shoreline.

2. Weathering on the coast

(a) *Salt crystallisation* is effective where evaporation potential is high and the growth of chloride salts derived from the salt in seawater attacks a wide variety of rocks. This type of rock destruction was discussed in Chapter 2 (page 21). Its main action is to loosen fragments for the processes of erosion.

(b) *Solution* on the coast is particularly important on limestones. Seawater is often already saturated with calcium carbonate and it is difficult to see how solution can be as effective as it is, particularly in the tropics, where the solubility of calcium carbonate decreases with rising temperatures. It is possible that photosynthesis in plants plays a part in the process by oxygenating the water during the hours of daylight while absorbing CO_2, and giving out CO_2 at night as a by-product of respiration. This raises the acidity of the water and so increases the solutional effect. The morphological effect of solution is to pro-

duce sharp fretted pinnacles of limestone called *lapiés*, low down in the intertidal zone (Fig. 9.11). Where wave attack is less prevalent, solution may cut
notches at the edge of pools and give small overhanging lips. In tropical seas
with microtidal or mesotidal ranges these features become very large and are
called *visors*.

Fig. 9.11 Coastal solution: a) solution landforms on a cool temperate coast and   b)
notch and visor on a tropical limestone coast (After Guicher) HWM high water mark;
LWM low water mark

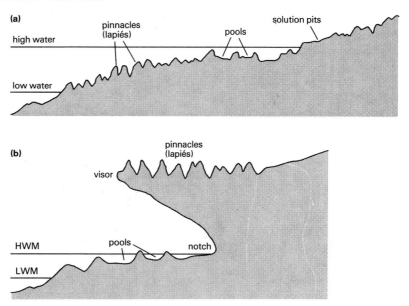

(c) *Biological activity* frequently assists solution (page 25). Other rock types
besides limestones are attacked by secretions, particularly from the blue-green
algae which live between the tide lines. Molluscs like the bivalve *lithophagus*
bore into the rocks of the shore in search of either food or shelter, and snails
searching for food wear down the rock surface with their radula or horny
tongue. Seaweed firmly attaches itself to rocks and in effect increases the surface area so that the rock is more easily moved by the waves; this is comparable to the way a sailing boat uses its sails to make maximum use of the wind.

3. Wave refraction

Figure 9.12 shows an indented coast with a uniform submarine slope along its
length. The wave crests are shown approaching the shore perpendicular to the
general trend of the coastline. The segments of the crests approaching the headlands begin to feel the sea bed when they are just under a kilometre from the
shore. They rise in height, decrease in wavelength and slow down (page 233).
The same crest approaching the bay continues unimpeded and so moves ahead
of the wave segment off the headland. Segments A and B at position 1 in Fig.
9.12 are in deep water and are unchanged. By the time they have reached
position 3, A has slowed down and shortened its wavelength. It therefore lags

Fig. 9.12 Wave refraction

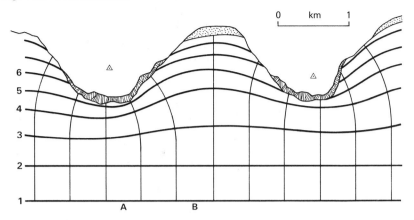

behind B which is still unchanged. By the time the wave reaches position 5, A is about to break on the headland while B is advancing more slowly into the bay. The end result is that the crests try to conform to the outline of the shore and to break parallel to it. Segments A and B in deep water were the same width. The *orthogonals* which are drawn at right angles to the crests from the ends of segments A and B show that the length of A is shortened by about 30 per cent at its break point, position 5, and B is lengthened to more than twice its deep water value. This means that the energy in segment A is concentrated onto the headland which causes wave height to rise in addition to the wave heightening caused by the shallowing of the water. Thus, since wave energy is proportional to wave height (page 232), the power of the waves is greater on the headland. In the bay wave height is less since the energy of segment B is spread out. The orthogonals drawn from the ends of the beach at the head of the bay show that there is a considerable decrease in energy from the deep water value.

The distortion of the wave crests is called *refraction* and where there is a shallow gradient down from the shore the wave approach tends to be nearly parallel with the shoreline and the crest breaks almost simultaneously along its length. If the slope is steep, the wave approach is not quite parallel and the break runs along the length of the crest. This is more evident in waves with a short wavelength since they are not influenced by the bottom until they are relatively near the shore.

4. Erosional landforms

(a) *Cliff form*

Sea cliffs can be regarded essentially as normal slopes with processes of the types described in Chapter 3 operating on them, and wave action providing the mechanism for basal removal. Frequently this is very efficient and all the material weathered from the cliff and falling to its base is removed (Fig. 9.13a). If an accumulation occurs, the base of the cliff is protected from wave attack. The energy of the waves is expended in attrition of the debris to a suitable size for transport. Mass wasting becomes the dominant process on the cliff and it

Fig. 9.13 Cliff form and structure

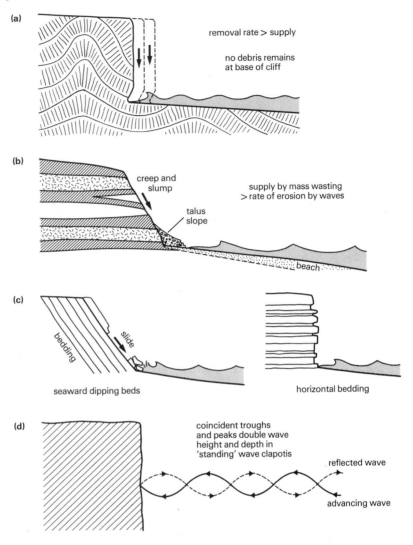

(a)

removal rate > supply

no debris remains
at base of cliff

(b)

creep and
slump

supply by mass wasting
> rate of erosion by waves

talus
slope

beach

(c)

bedding

slide

seaward dipping beds

horizontal bedding

(d)

coincident troughs
and peaks double wave
height and depth in
'standing' wave clapotis

reflected wave

advancing wave

becomes less steep (page 67). The relationship between mass wasting and basal removal is critical in determining the form of the cliff. If there is an abundance of wave energy, debris will be removed and the foot of the cliff will be exposed to active abrasion. In macrotidal environments (page 236) like the British coast this results in an undercutting of the cliff at about the high tide line. Such undercuts are called *wave cut notches* and are recognisable as abrasion features by the smoothed nature of the rock. Weaknesses in the rock are picked out and erosion proceeds along these at a faster rate, forming *sea caves*. In weak or unconsolidated rock coasts, these features do not form and slumps and rock slides occur as the base of the cliff is undermined (Fig. 9.13b and c and page 51). The best developed wave cut notches are found in storm wave environments in resistant rocks with microtidal ranges, which concentrate wave energy at one height on the shore.

241

(i) *Plunging cliffs* descend directly into deep water and as a result the largely unaltered ocean wave hits them directly. The changes associated with the breaking wave do not occur and the wave energy is not used to do work but is reflected away from the cliff at the same angle as its angle of incidence. The advancing wave and the reflected wave interfere to give a standing wave called a *clapotis*. This makes it appear that waves do not advance but cause the water surface to oscillate about nodes (places where the surface appears static) one wavelength apart (Fig. 9.13d).

(ii) *Slope over wall cliffs*   In parts of Britain the effect of two phases in the history of coastal cliffs can be seen. In Plate 9.1 the cliff just above sea level is the product of present day wave action. The more gentle gradient higher up, the 'slope', was formed when the climate was colder and sea level much lower as was explained in Chapter 8. Periglacial conditions resulted in the degrading of the cliff and wave action, on the return of the sea to its present level has removed the debris from its foot and resumed its attack. Such cliffs are found around the coast of Devon and Cornwall and in Wales, outside the most recently glaciated area.

Plate 9.1 Slope over wall cliffs near Fishguard, Dyfed (Author's photograph)

(b) *Lithology and structure*

On most coasts lithology and structure exert a very strong control on both the profile and the plan of the cliffs.

(i) *Profile*

Where rocks dip seawards, the cliff develops as a long slab (Fig. 9.13c) and rock slides are the usual form of mass movement. Horizontally bedded and

242

inland dipping strata usually produce steeper cliffs, since in hard rocks the strata form ledges which hold material on the cliff.

(ii) *Plan*

In plan the general tectonic trend influences the outline of the coast. Where the trend of folds is at right angles to the trend of the coast, a series of bays and headlands develops. The area of south-west Ireland (Fig. 9.14a) shows this type of '*Atlantic*' coastline. If the general trend of the fold structure is parallel to the coast it is called a '*Dalmatian*' coastline after an area of Yugoslavia near Split. This results in elongated islands and inlets running parallel to the coast (Fig. 9.14b).

Fig. 9.14 Coasts in plan

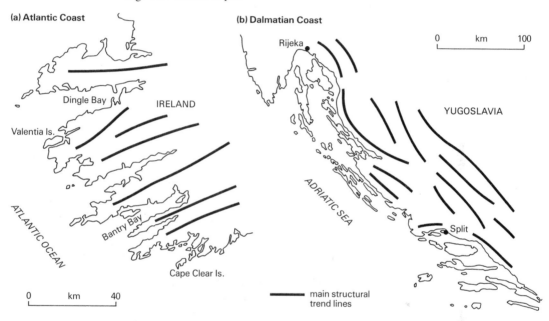

(iii) *Bays and headlands*

On a much smaller scale, differing rock hardness is picked out and exaggerated by wave action. Hard rocks form headlands and the softer rocks are eroded into bays. The differences in resistance may be quite small, but long periods of wave attack emphasise them. The south coast of England between Bournemouth and Weymouth is one of the areas in Britain where the effect of lithology on the landscape is seen to best effect. Figure 9.15 shows the geology of the area around Lulworth Cove. The structures run parallel to the coast. The harder Portland Beds form the southernmost part of the coast. In places the sea has breached this hard band and been able to erode the softer rocks, the Wealden Beds, which are adjacent to them to the north. This is apparent at Lulworth Cove itself where the entrance in the Portland Beds is still narrow, while inside the cove wave refraction has fanned the waves out and enlarged the cove into an almost circular shape. This now impinges on the Chalk, a relatively

Fig. 9.15 The geology of the coast between Lulworth Cove and the Isle of Wight

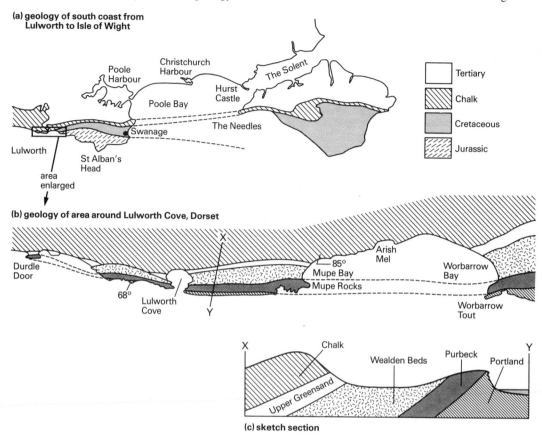

(a) geology of south coast from Lulworth to Isle of Wight

Poole Harbour
Christchurch Harbour
The Solent
Hurst Castle
Poole Bay
The Needles
Swanage
Lulworth
St Alban's Head
area enlarged

Tertiary
Chalk
Cretaceous
Jurassic

(b) geology of area around Lulworth Cove, Dorset

Durdle Door
68°
Lulworth Cove
X
85°
Mupe Bay
Mupe Rocks
Y
Arish Mel
Worbarrow Bay
Worbarrow Tout

(c) sketch section

X
Chalk
Wealden Beds
Purbeck
Portland
Y
Upper Greensand

harder rock. The cove will probably develop in the same way as Worbarrow Bay 2 km to the east. Here the bay has been enlarged in an east-west direction with Warbarrow Tout and Mupe Rocks marking the position of the Portland Beds and forming headlands. The northward expansion of the bay has been prevented by the harder Chalk although at Arish Mell a pre-existing valley through the Chalk ridge provides a low point through which the sea may cut to erode a wider bay in the Tertiary rocks to the north. The Chalk ridge to the east of Swanage has been breached completely and the sea is actively attacking the Tertiary beds between Bournemouth and the Solent. The backbone of the Isle of Wight is the continuation of the Chalk.

(c) Shore platforms

(i) Many cliffs have an area of gently sloping rock at their foot which may extend seawards for up to 1 km and occasionally even further. These *shore platforms* frequently cut across different lithologies and structures with little alteration in form. Around the British coasts, in a storm wave, macrotidal environment, they usually slope gently between the high and low tide limits and link on their landward side to the wave cut notch. These are *intertidal shore*

244

platforms and they are evidently abrasional features, being frequently covered with a veneer of pebble, cobble and boulder sized debris derived from the erosion of the cliff. Plate 9.3 shows a platform on the south coast of England covered with chalk boulders from the high cliffs in the background. There is evidence of cliff falls which indicates that the base is being actively attacked by the waves. The wave cut notch, while bringing about the erosion of the cliff (page 238), is extending the platform landwards. This is shown diagrammatically in Fig. 9.16a. The wave cut notch is extended at high tide level and the abraded debris is transported in the breaker zone. The simple notion that this debris goes to form a terrace just beyond the edge of the platform cannot be substantiated since such terrace features are of rare occurrence. The material is either carried out into the deeper offshore zone where wave action is absent, or it is moved laterally along the shore to areas where waves are less active to form beaches (page 248).

The gradient of intertidal platforms varies with changes in lithology and structure, fetch, orientation, wave exposure and type. At Eastbourne it is about 0.5° and at Studland near Swanage (Fig. 9.15) it is about 4°. Both these areas are in Chalk. In detail, the platforms may show a stepped profile in which level

245

Plate 9.3 Chalk cliffs and the shore platform near Dover (Derek Widdicombe)

sections represent pools. This is particularly noticeable in rocks which are prone to chemical or biochemical attack (page 25).

As the cliff retreats and the platform grows wider, the waves dissipate their energy over an increasing distance. There is much less erosion of the cliff and the wave cut notch may cease to be eroded. The energy of the waves is used in a number of ways, two of which are relevant here. First, the size of the debris on the platform is reduced by attrition, and second, the platform itself is lowered by abrasion (Fig. 9.16a). If the platform is lowered significantly, the

Fig. 9.16 Types of shore platforms

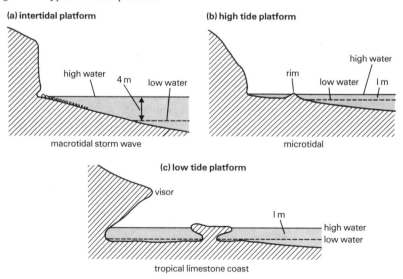

(a) intertidal platform

high water   4 m   low water

macrotidal storm wave

(b) high tide platform

rim   high water   low water   I m

microtidal

(c) low tide platform

visor   I m   high water   low water

tropical limestone coast

waves resume their attack on the cliff. This indicates how closely cliff recession and platform abrasion are linked, since if the cliff supplies more debris, platform abrasion is slowed down.

The width of intertidal platforms has posed something of a problem, many are far wider than they should be according to studies of the depth of wave erosion. In California the surface of sand grains indicated that wave erosion was ineffective below a depth of approximately 10 m. If the platform slopes at 1°, the outer edge of the platform lies 800 m from the high tide shoreline if the tidal range is 5 m, which places it in the macrotidal group. Many platforms, including that in Plate 9.3, are wider than this and are found in areas with a lesser tidal range. The solution to this problem of 'overwide' platforms lies in the recent history of sea level (page 269). Over the last 18 000 years sea level has risen almost continuously. The platforms thus grew steadily with the rising sea.

(ii) *High tide platforms* are found at about mean high tide level in microtidal swell wave environments as far apart as Tasmania and Turkey. They are remarkably horizontal and are caused by processes other than abrasion. Sometimes their outer edge has a rim where it is continually wetted by wave splash (Fig. 9.16b). This gives a vital clue to their origin. The permanently wet zone is weathered less than the area which is sometimes dry, which indicates the operation of salt crystallisation or wetting and drying on the platform. This is probably assisted by biological activity which is localised at a particular level (page 239). Infrequent wave action on these platforms removes the weathered material and exposes fresh rock to these weathering processes. In areas more exposed to wave action, abrasion predominates and these processes are relatively unimportant.

(iii) *Low tide platforms* are found mainly in tropical seas where the coast consists of calcareous rock. Biological weathering and wave action combine, producing a platform backed by a visor (page 239) at the level of most intense biological activity (Fig. 9.16c).

ASSIGNMENTS

1. *Distinguish between weathering and erosion. Compare the effects of weathering in coastal areas with those in inland areas. Indicate which methods of weathering described in Chapter 2 are important or unimportant on the coast.*

2. *On the basis of the principles described and explained on page 239, draw the wave refraction patterns which you might expect to occur if waves travelling from south to north met: (i) a small circular island, (ii) a narrow peninsula projecting eastwards from a land mass, and (iii) a narrow south facing harbour mouth with two breakwaters protecting the harbour. Which area of each would be least prone to wave attack?*

3. *A relatively uniform soft rock area becomes rapidly submerged by the sea to form a 'new' coast. It lies in a macrotidal storm wave environment with the maximum fetch about 1000 km with the prevailing wind from the south-west. Describe the processes operating on such a coast to produce cliffs and shore platforms. What would happen when the platform became so wide as to limit wave attack on the cliffs?*

4. For each of the three types of shore platform, list the dominant processes. Explain the reasons for their differences in form.

C. Beaches

In Britain most beaches are composed of sand or shingle with varying proportions of plastic bottles, tin cans, bricks, shells and drift wood. In some cases the beach may be made up entirely of one material. On Brownsea Island in Poole Harbour, Dorset, for example, there is a beach made completely from broken sewer pipes. In Canada there are beaches of timber, and in the USA of tin cans. The Outer Hebrides are fringed along their west coast by the machair, dunes derived from beaches of shell fragments. Calcareous beaches are more typical of tropical than temperate coasts. If we define a beach as an accumulation of material at about the high tide shoreline, then all the above are legitimately beaches. However, most beaches are composed of the products of rock breakdown with varying proportions of biological material, usually shell fragments.

1. Sources of beach material

The reasons for the accumulation of beaches fall into two simple groups. A beach may be a *store* of sediment trapped in an embayment on the coast, or it may be a constantly mobile *stream* of sediment which represents a transport pathway for sediment moving along the coast. Ultimately this mobile sediment becomes trapped and forms a beach in which there is a more limited movement.

The sediment which makes the beach has four main sources.

(i) *Material eroded from headlands* by contemporary wave attack is usually only important where the rock is soft or unconsolidated. Some headlands in hard rocks show no appreciable modification after 5000 years of wave attack. The chalk cliffs of southern England over about the same time period have retreated hundreds of metres in places but the flint which forms the beaches in the area must have been derived from a much greater volume of chalk than this retreat represents.

(ii) Sediment may be moved onto beaches from the *offshore zone*. We have seen that wave action has little effect below 10 m, which means that there is a relatively small area in the offshore zone from which beach material may be derived. In places where the gradient is steep the 10 m line at low tide is very close to the shore, yet there are massive accumulations of beach material. Chesil Beach (Plate 9.6) is a good example of this. However, sea level has risen more than 100 m in the past 18 000 years, and it is probable that the advancing sea swept up large volumes of material from the continental shelves to form beaches like Chesil. In areas which have been glaciated large volumes of deposits have been dumped on the continental shelf when sea level was low (page 270). This soft, unconsolidated sediment was eroded and reworked by the sea and probably forms a large part of the beaches in such areas.

(iii) *Large rivers* bring considerable volumes of sediment into the coastal zone. Frequently this is of silts and clays and is carried beyond the zone of wave action. The energy of the waves would be too great to allow it to settle on the beaches. Sand and coarser material are carried only by turbulent rivers and this source of sediment for beaches is restricted to the wetter parts of the world. The Rhine, swollen by the meltwaters of the last glaciation, brought large volumes of coarse debris to the coast of the North Sea. The outwash rivers of southern Iceland and South Island, New Zealand, have provided much of the gravel in the shingle features of those coasts.

(iv) Beach material cycled through one beach and then *moved along the coast* to the next beach, although a component in any one beach must initially have come from one of the other sources.

2. The nature of beach materials

The majority of beaches are composed of either sand-sized material, 1/16 mm to 2 mm, or of pebbles, 4 mm to 64 mm and coarser. The granule group, 2 mm to 4 mm, is poorly represented in beaches as a whole. It is also unusual to find a beach composed of a mixture of sand and pebbles. Plate 9.4 shows the beach at Newgale, Pembroke, where the waves have sorted the pebbles into a steep shingle ridge, while the sand forms the gently shelving foreshore. There is little granule-sized material evident.

Plate 9.4 A mixed sand and shingle beach, Newgale, Pembroke (Author's photograph)

Sand beaches

The sand of most beaches is usually very well sorted; a small range of particle sizes is present, frequently less than 1 ø (page 9), although at different places on the beach, sand with a different mean size may be found. The results of a grain size analysis of a beach sand is shown in Fig. 9.17a as a cumulative frequency curve. The median size is 1.5 ø and the interquartile range is 1.0 ø. The histogram compiled from the same data shows that the mean is displaced towards the coarse end of the distribution in relation to the median. The dis-

Fig. 9.17 Grain size distributions in beach sands

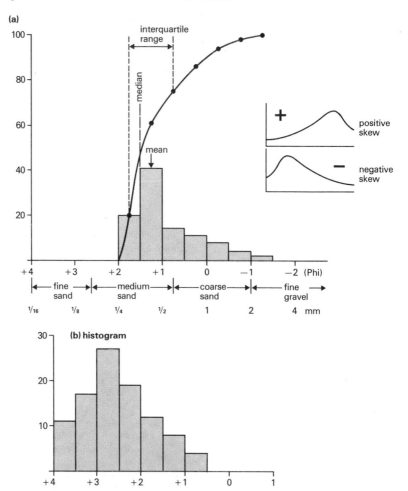

tribution is *negatively skewed* and has a tail of coarse particles. This characteristic of beach sand is due to the waves selectively removing the finer sizes and it is useful in helping to distinguish wave deposited sediments from other environments of deposition such as dunes and rivers.

The high degree of sorting found in beach sediments is due to the differential effect of the swash and backwash. If the wave is of the spilling type, the coarser material is moved up the beach by the strong swash. The backwash is weaker as some of the water returns by *percolating* down through the sand. The coarsest particles thus tend to be found at the limit of wave action while the finer ones are moved into the lower energy zone offshore.

The *size* of the beach material has an effect on the *slope* of the beach in the *swash zone or beach face*. Plate 9.4 shows how two different slope angles are developed on sand and shingle. In general, the coarser the sediment the steeper the slope. Table 9.1 relates slope angle to particle size but this is a simplified situation; more exposed beaches tend to have lower slope angles and sheltered beaches tend to be steeper than the table suggests.

Table 9.1 Beach slope and particle size

Cobbles	24°
Pebbles	17°
Granules	11°
Very coarse sand	9°
Coarse sand	7°
Medium sand	5°
Fine sand	3°
Very fine sand	1°

3. Beach profiles

Beach profiles may be regarded in the same way as slope profiles (Chapter 3) in that they reflect the vertical variation in processes down the beach. Figure 9.18 indicates some of the major zones of the beach profile. The *berm* is a horizontal

Fig. 9.18 Sediments and landforms on a sand beach

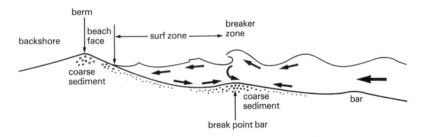

or gently inclined area at the top of the beach which slopes seawards and experiences wave action infrequently. Usually there is a sharp break of slope to the steeper *beach face* which is the zone affected at high tide by wave action. Sand beaches are gently sloping and the finest appear to be almost flat. Slopes of ½° are not uncommon. In such a situation the waves break some distance from the shoreline at the *break point bar* and run towards the beach as *surf*.

The profile of a beach can change very rapidly. The violent storms of the night of 31 January 1953 (page 282) moved large volumes of sand from the beaches of the North Sea coast and in cases lowered the level of the beach by more than 1 m in one tidal cycle. Storm waves tend to erode beaches very quickly, while swell waves build the profile back up slowly. Figure 9.19 shows how a beach profile may vary. The *summer profile* is usually higher than the *winter profile* since storm waves are less frequent and wave action tends to be constructive. The zone of change between the two profiles is termed the *sweep zone*.

(a) *Erosion and deposition by waves*

Whether a wave exerts a constructive or an erosive action on the beach is dependent on a complex interaction between the wave and the beach itself. The

Fig. 9.19 Beach profile changes

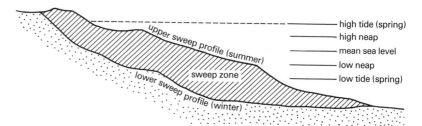

most important wave parameter in determining this is wave steepness (page 232). A deep water steepness ratio of 0.025 is the approximate dividing line between the less steep constructive wave and the steeper destructive wave. Other conditions on the beach may alter this and waves with low steepness values may be seen to be erosive and steeper waves to be constructive.

The wave period determines the relative dominance of the backwash. If the wave breaks just as the backwash of the preceding wave is running down the beach towards the breakpoint, the swash has a limited effect and erosion by the backwash is dominant. If the backwash has subsided before the next wave breaks, the swash is highly effective and construction is the result. W.V. Lewis in 1931 determined that destructive waves had periods of 4–5 seconds while constructive waves had 8–10 second periods.

The direction of the wind in relation to the coast can cause considerable modification of the effect of the waves on the beach. An onshore wind tends to give steep waves which are generally erosive and it also produces a drift of water towards the shore at the surface and a compensating return flow along the bottom. This flow is relatively weak but is capable of moving sediment already in suspension in the water. The transportation of sediment requires far less energy than erosion of the same sized particles (page 107). The onshore wind accentuates erosion. It is usually this wind that does most of the work in moving sediment and is thus described as the *dominant wind*. In the west of Britain the dominant wind direction is frequently the same as that of the *prevailing wind*, that is from the west or south-west. In contrast, on the east coast the prevailing wind is offshore and accomplishes little work. Onshore winds from the north-east or east produce the greatest change in the coastal zone, since these have the greatest fetch and generate the largest waves (page 230).

(i) *Break point bars*

Where steep waves are breaking on the shore with a frequency which allows the backwash to move material down the beach, an accumulation occurs at the break point, since movement along the bottom, outside the break point, is shoreward (Fig. 9.18). This forms a *break point bar*. In a microtidal sea such as the Mediterranean these features form where the waves break for the longest period of time, which is at the low tide and high tide break points. In a sea with a larger tidal range, a bar formed by a storm at low tide will not be affected by waves of average height at other states of the tide since it will be below the depth of effective wave action. Bars formed higher up the beach will form less permanent features since they will be affected by smaller waves as the tide falls.

On beaches where long swell waves are the norm (Fig. 9.21), the strong swash builds up a bar at the high tide shoreline. These *swash bars* are not normally found on British coasts where storm waves and shingle preclude their formation.

(ii) *Ridge and runnel beaches*

In macrotidal environments with limited fetch (page 230) the sand beach often becomes fashioned into a series of large undulations 100–200 m from crest to crest and up to 1 m in height. Plate 9.5 shows these on the south Lancashire coast near Formby. The crest of the *ridges* tends to be orientated perpendicular

Plate 9.5 Aerial view of ridge and runnel beach at Ainsdale, Lancashire (Cambridge University Collection — copyright reserved)

to the maximum fetch and where the coast changes direction the ridges continue their trend so that they run at an angle to the coast. The depressions between the ridges are called *runnels* and water collects in these until it cuts a small channel through the ridge at a low point as the tide falls.

(iii) *Profiles of equilibrium*

The unconsolidated nature of beach sands means that the beach responds rapidly to changes in the wave conditions. If the beach is steep, the waves tend to be of the destructive type since they plunge and comb down the beach. This will result in the lowering of the beach gradient so that the wave break becomes more gradual and the waves start to spill and become constructive, raising the beach steepness.

It follows that there is a point between these two states when the profile is being neither raised or lowered. It is known as the profile of equilibrium. It is comparable to the idea of grade in river systems (page 121), using feedback between the variables to regulate the system.

In reality it is very difficult to identify the equilibrium profile. Each wave differs from the preceding one and the waves arrive in *trains* so that the equilibrium profile constantly changes. Prolonged periods of fairly constant swell waves tend to produce a situation approaching that of the theoretical equilibrium profile but the beach responds rapidly when the waves change.

(b) *Shingle beach profiles*

Shingle beaches often show marked ridges in the profile (Fig. 9.20). These are *storm ridges*, produced by storm waves which hurl shingle above the limit of wave action. At high tide the waves break at one place on the beach while the

Fig. 9.20 The profile of a shingle beach

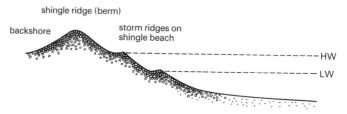

tide turns. If the successive tides are lower, as during the period after the spring tides, then each high tide forms a ridge lower down the beach. The majority of storms rarely last more than two or three tidal cycles, and winds of less than Beaufort force 4 produce waves whose effect on the beach is very limited. The end result is usually one storm ridge produced per storm when the wind is onshore. Chesil Beach (Plate 9.6) and Newgale beach in Pembroke (Plate 9.4) have both been built up into a high ridge which falls sharply on the landward side. Chesil, at its eastern end, is more than 14 m above mean high tide level.

The clear segregation of sand and shingle was explained on page 249. Because of this segregation, shingle beaches show a pronounced difference in slope angle from sand beaches (page 251). Chesil, at its eastern end where the pebbles are about 50 mm in diameter, has a slope of 1 in 2 or 26°. When waves are breaking on beaches like this it is possible to hear the backwash dragging the shingle down the beach. The percolation of water through the beach is dependent not only on particle size but also on the saturation level of the beach. If water is present near the beach surface in a shingle beach, percolation is less rapid and the backwash is much stronger. Compare this with the conditions which give rise to overland flow (page 79). In this way the water table plays a very important role in controlling the effect of the swash and backwash.

Plate 9.6 Chesil Beach from the west (Aerofilms)

4. Beaches in plan

(a) *Swash aligned beaches*

On maps the most striking feature of all beaches is their smoothly curving outline. Generally they are concave towards the sea. Convex beaches do exist under special circumstances but are rare (page 262). This concavity is due to the effect of wave refraction (page 239). The beach face tends to orientate parallel to the wave fronts. This is best illustrated in swell wave environments where the wave advance is regular and the beach less prone to adjustment to the individual storm. Storm Bay, Tasmania, shown in Fig. 9.21 illustrates almost perfect adjustment to the refracted wave fronts. In storm wave environments the beaches tend to orientate themselves parallel to the fronts of the dominant waves. The beaches of Morfa Harlech and Morfa Dyffryn at the head of Cardigan Bay on the coast of Wales (Fig. 9.22) have deviated from the line of the coast and face more towards the direction of maximum fetch which is southwest towards the South Atlantic, a fetch distance of more than 9000 km.

At Santa Monica, California, a breakwater was built parallel to the shore in order to protect anchorages near the pier. Wave refraction around the breakwater caused the originally straight beach to change its plan to face the

Fig. 9.21 Beaches and wave fronts in Storm Bay, Tasmania (After Davis)

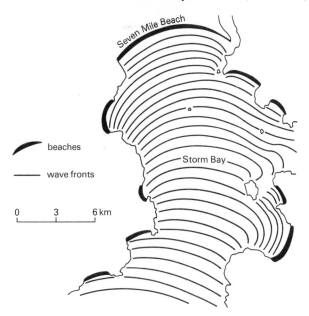

refracted fronts. This building out of the beach reduced the area available for boats and caused the shallowing of the water around the pier (Fig. 9.23).

Beaches which face the wave fronts are termed *swash aligned* and have a limited movement of sediment along the beach. The orientation of Chesil Beach is almost at right angles to the direction of maximum fetch and experi-

Fig. 9.22 Morfa Harlech and Morfa Dyffryn, Wales

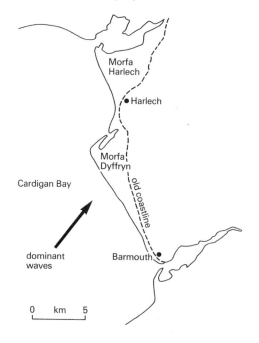

256

Fig. 9.23 The effect of a breakwater at Santa Monica, California

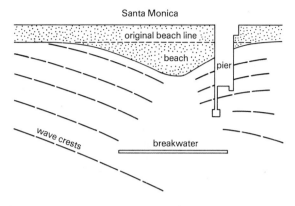

ments using marked pebbles have shown that there is little net movement of material along the beach (page 285).

(b) *Drift aligned beaches*

A large number of beaches show alignments at an angle to the dominant wave fronts. This usually occurs where the beach gradient is steep and the wavelength short so that there is insufficient time for the wave to become refracted perfectly between feeling the sea bed and breaking on the beach (page 240). As a result the wave crest breaks at different times along the beach. In the swash zone this results in a particle being washed up the beach diagonally (Fig. 9.24) by the swash which slows towards the top of its run and changes to the backwash which runs down the steepest line of the beach. Since the backwash tends to be weaker than the swash, the particle will tend to move up the beach until a larger wave drags it back down, as in wave 3, Fig. 9.24. The slight movement along the beach happens to millions of particles at the surface of the beach as the wave breaks. In 24 hours, 15 000 waves may break on the beach and the total movement of sediment may be very great. This process is called *longshore drifting* or *beach drifting*.

Fig. 9.24 Beach drifting

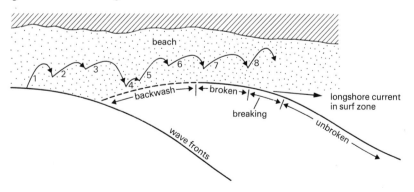

In the surf zone the diagonal breaking of the waves induces a movement of water along the shore called a *longshore current* which, although usually weak, is capable of transporting sediment which is already in suspension in the water, as was explained on page 000.

(c) *Changes in beach alignment*

The longshore transport of sediment by these processes can determine the orientation of beaches, particularly where there is little to halt the flow of sediment. The most effective direction of wave advance for producing longshore sediment movement is between 40 and 50° (Fig. 9.25a). If the angle of wave advance becomes less than this (Fig. 9.25b), all the sediment in transport can no longer

Fig. 9.25 Swash and drift alingments

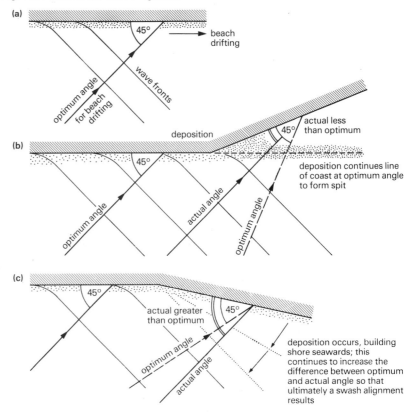

be moved and deposition occurs. The accumulation of beach material continues until the angle between the beach and the direction of wave advance is narrowed to the optimum of 40–50°. The usual result of this is that a spit is built out from the point where the coast changes its orientation. The spit continues the line of the best orientation for causing longshore drift in relation to the most effective waves.

If the coast turns seawards so as to increase the angle of wave approach to the coast (Fig. 9.25c), deposition again occurs so that the angle continues to become greater. The limiting case of this is where a swash alignment is achieved and accretion builds the beach seawards at a constant angle.

258

5. Small scale landforms on beaches

(a) *Beach cusps*

When a steady swell approaches a beach with almost perfect refraction, a series of small arcuate embayments between 5 and 50 m across sometimes form. These are *beach cusps* and on long beaches hundreds may lie in a line parallel to the water's edge (Plate 9.6). Their origin is difficult to explain. It is easier to see how they are perpetuated. The sides of the cusp channel the swash into the centre of the arc. The backwash in the centre is far stronger as a result and combs down the beach, deepening the cusp. The sediment in the 'horns' tends to be coarser than in the cusp itself. Cusps tend to form more frequently on coarse sand and shingle beaches. They are bigger when the swell is higher.

(b) *Rip channels*

On many beaches where wave advance is adjusted to the beach plan, the swash piles water up in the surf zone. The water tries to escape laterally (Fig. 9.26), but is opposed by water attempting to escape in the opposite direction.

Fig. 9.26 Water movements on a surf beach

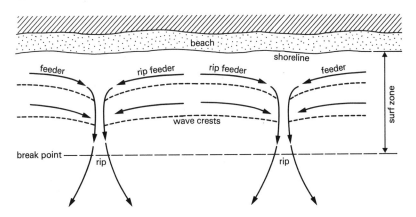

The two streams together have sufficient energy to break out through the surf as a well defined zone of high velocity seaward moving water. The velocity of this *rip current* in cases may be as high as 8 km/h and can scour temporary channels in the sand. Rip currents can be observed moving sand out beyond the breakers. They are more common on swell wave beaches.

(c) *Ripples*

A wave in shallow water (page 232) oscillates sand on the seabed back and forth. This moves sand into small ridges elongated in the direction of the wave crest. These *ripples* are symmetrical and are up to 100 mm high and 500 mm from crest to crest (Fig. 9.27).

Asymmetrical ripples are produced when there is a strong directional flow. The current moves particles up the gently sloping up-current side and then

Plate 9.7 Ripples on a sand
beach (Author's photograph)

Fig. 9.27 Ripples on a beach

(a) symmetrical wave ripple

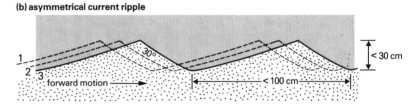

(b) asymmetrical current ripple

dumps them in the eddy zone in the lee of the ripple, where they lie at the angle
of rest for that particular size of particle. In rip channels and in estuaries where
the currents are particularly strong these current ripples may be 0.4 m high and
about 4 m long.

ASSIGNMENTS

*1. Compile a cumulative frequency curve from the histogram in Fig. 9.17b.
   What is the modal class and interquartile range for this deposit? Using
   descriptive terms for the size classes (page 9) describe this deposit.
   Account for the differences between (a) and (b) in Fig. 9.17.*

260

2. Using Fig. 4.20, explain how sand and shingle on beaches become segregated.

3. A beach profile surveyed in legs of 2 m length produced the following results in a student's notebook:

> Edge of lagoon behind bar (47 m south of lifebelt post) Bearing 264° magnetic
>
> +2°, +4°, +7°, +1°, 0, 0, −4°, −12°, −11°, −13°, −14°, +2°, −12°, −14°, −11°, 0, 0, −14°, −11°, −11°, −13°, −5°, −4°, −3°, −4°, −3°, −4°, 0, +2°, +1°, +3°, 0, +1°, −4°, −2°, −3°, −2°, −2°, +1°, −2°, waters edge (low tide).

(a) Construct the beach profile. Label berm, bars, shingle, sand and storm ridges. What type of wave and tide environment might such a beach be found in?

(b) Explain where and how beach cusps and ripples could be formed on the profile.

4. A storm beach on the north coast of Scotland orientated east-west contained four dominant lithologies: gneiss, arkose (a sandstone containing feldspar), granite and quartz from veins. At four equally spaced points along three kilometres of the beach a sample of 50 pebbles was measured and the proportion of each rock type counted. Table 9.2 shows these results.

Table 9.2

	Gneiss	Arkose	Granite	Quartz	Mean Size (mm)
Station 1	18	22	7	3	18
Station 2	13	17	12	8	11
Station 3	13	14	13	10	10
Station 4	12	12	17	9	11

Which is the most likely direction of drift? Draw a divided bar graph for each location to indicate the proportion of each rock type. Explain which are the most resistant and least resistant rock types.

5 A profile of equilibrium was defined on page 254 as a feedback mechanism comparable to that of grade in rivers (page 136). Using beach slope, erosion and wave type (regarding destructive as negative and constructive as positive), construct a diagram similar to that on page 121. Explain which of these variables can be negatively correlated and which positively. What is the negative feedback which regulates the system?

6. Explain the differences between swash and drift aligned beaches. A section of coast has a trend of 315° for 20 km and then turns to trend 270° for 20 km. Label areas of the coast where swash and drift alignments are likely to occur if the deepwater wave fronts are aligned east-west. Where would accumulations occur to form major landforms? If the wave fronts changed to be aligned along 295°, in which direction would drift occur along this coast?

D. Constructional Landforms

The variety of forms which deposition takes are illustrated diagrammatically in Fig. 9.28. They are named on the basis of their shape rather than their origin.

Fig. 9.28 Depositional features of the coast (After King)

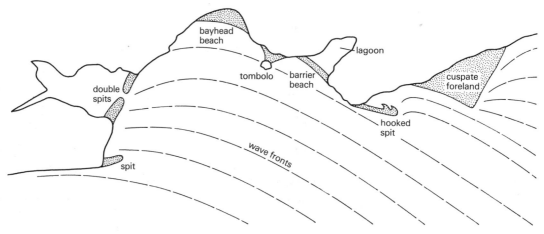

1. Bayhead beaches

This is the simplest type of beach and is a sediment trap into which material drifts from both directions. Bayhead beaches were described in relation to swash alignments on page 255.

2. Spits

Spits are amongst the most common of the major depositional features. Spurn Head shown in Plate 9.8 is a long spit which continues the line of the Holderness coast of Yorkshire. It is drift aligned and receives sediment drifting south from as far to the north as Bridlington. To the north of this, the headland of Flamborough Head forms a barrier to beach drifting from further north. The rapid erosion of the glacial deposits of this lowland coast supplies abundant sediment to feed the spit. The dominant waves arrive from the north and north-east and full refraction does not occur. This produces a strong drift to the south which has built out the gently curving sand spit into the deeper water of the Humber estuary. In the quieter water behind the spit finer sediment has been deposited. The Holderness coast has retreated about 4 km since Roman times and the sediment eroded has considerably lengthened the spit (Fig. 9.29). The *proximal end* of the spit, the end attached to the land, must also have retreated by this amount.

The growth of spits by longshore drifting across river mouths frequently forces the river to the downdrift side of the estuary. The Humber has been diverted to the south by Spurn Head. Similarly in Suffolk, the River Alde has been diverted 11 km to the south by the growth of the shingle spit of Orford Ness.

Fig. 9.29 Development of the Holderness coast and Spurn Head (After Pitty)

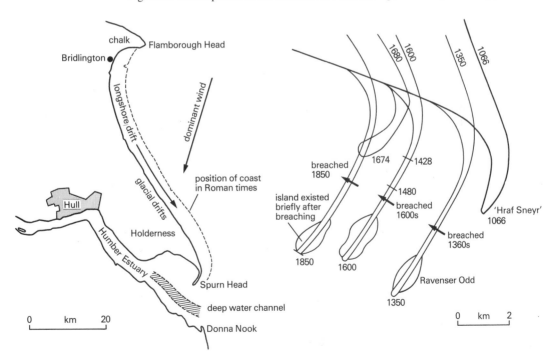

263

(a) Hooked or recurved spits

As spits build out into deep water they require increasing volumes of sediment to build up above the high water mark. This has the effect of turning the tip or *distal* end of the spit towards the land as the waves in the deeper water have a greater energy to attack the spit. Hurst Castle spit has built out from the coast of Hampshire into The Solent and has a number of hooks at its distal end (Fig. 9.30). The curves are so excessive that they could not have been produced by one direction of wave advance. The dominant waves arrive via the Channel from the south-west. Occasionally, however, winds blowing from the east and south-east produce drift-forming waves down the Solent (Fig. 9.15). The distal end of the spit turns to face these waves. Once formed, these recurves are sheltered from the dominant waves by the spit itself and thus become permanent features. The recurves further down the spit towards the proximal end are related to earlier phases of spit formation when the coast stood seaward of its present position, as is illustrated in Fig. 9.30.

Fig. 9.30 Hurst Castle spit (After Lewis)

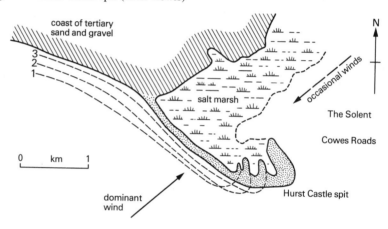

In swell wave environments where local winds above Force 4 are rare, and the swell generated in distant waters approaches the coast at a constant angle, recurves are uncommon. This lends support to the idea that they are caused by occasional waves from directions other than the dominant one.

(b) Double spits

Poole Harbour and Christchurch Harbour, both in Bournemouth Bay, have spits at their entrances which apparently have grown into the embayments from opposite directions. The history of these features over the past 300 years is well known and helps explain their apparently anomalous drift directions.

The South Haven spit in Poole Harbour (Fig. 9.31a) in 1785 had a channel near its present proximal end at the position of the Little Sea, which was the main entrance to Poole. By 1875 the entrance had moved to its present position where it has since been held by training walls. The change of position was caused by a bar feature which formed offshore and was pushed towards the beach, thus blocking the exit. At Christchurch in 1880, a spit extended from

Fig. 9.31 Double spits

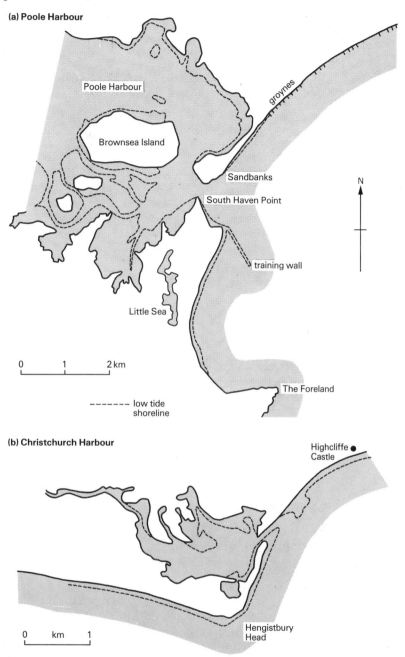

(a) Poole Harbour

Poole Harbour

Brownsea Island

groynes

Sandbanks

South Haven Point

training wall

N

Little Sea

0   1   2 km

The Foreland

------- low tide
shoreline

(b) Christchurch Harbour

Highcliffe
Castle

0   km   1

Hengistbury
Head

the south side of the harbour almost as far as Highcliffe Castle, diverting the entrance in the downdrift direction. The spit was breached in 1886 and again in 1935, in each case by a easterly storm (Fig. 9.31b). There is no evidence of a drift from the north forming the northern spit, and it is possible that this is the remnant of a former spit from the south which grew completely cross the harbour mouth before historical records were kept.

The two cases show attributes which indicate that they are swash aligned and drift aligned respectively. Christchurch is more exposed to the prevailing wind from the south-west. Thus drift has played a more important part in its formation than in Poole Harbour which is protected by St Albans Head (Fig. 9.15). The greatest fetch from Poole is 250 km towards Dieppe, and swell refracted round the headland probably contributes to the swell orientation of the spits towards the east-south-east.

3. Forelands

Dungeness is a huge area of shingle ridges forming a low-lying triangular area between Hastings and Dover. The pattern of the shingle ridges is shown in Fig. 9.32, and using this and historical information the history of the foreland has been deduced. The first stage was the growth of a spit from Fairlight Head which was seaward of its present position A. This may have grown right across

Fig. 9.32 Stages in the growth of Dungeness, a cuspate foreland (After Lewis)

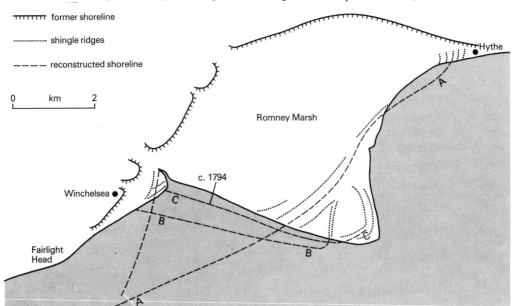

to Hythe by the thirteenth century. The supply of shingle from the west then apparently diminished and the proximal end of the spit was eroded and the shingle redeposited at the distal end, which was recurved. The spit thus rotated to face the dominant waves in positions B and C, a swash alignment. Infill in the lee of the spit became separated from the sea to form Romney Marsh as shingle drifted round the end of the spit, building the parallel ridges on the east side.

Another foreland is the area known as Benacre Ness in Suffolk near Southwold. Historical records show that the feature has moved northwards against the direction of drift which is from the north (Fig. 9.33). The updrift side traps sediment and slowly builds forward while the downdrift side is eroded, prod-

266

Fig. 9.33 The movement of Benacre Ness in historic time (After Williams)

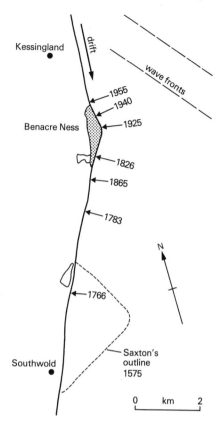

ucing a net updrift movement. In this case there appears to be an ample sediment supply.

These two types of foreland formation are represented by (a) and (b) in Fig. 9.34. A third type, already described in part is represented by (c). This type has two swash aligned faces and usually forms where an island protects the coast, reducing wave energy and refracting the waves approaching the shore. Frequently this type of foreland grows to link the island to the shore when it is called a *tombolo*.

4. Barrier beaches

A beach which stretches from one side of an embayment to the other and encloses a lagoon behind it is a *barrier beach*. Loe Bar near Porthleven is a shingle barrier, behind which there is a freshwater lagoon, The Loe. This feature is swash aligned and has probably been driven landwards from the offshore zone (Fig. 9.35). The east coast of the USA from Texas to Virginia is intermittently fringed by barrier beaches. On one of these the resort of Palm Beach has been built. These huge beaches have slowly been driven shorewards during a period of rising sea levels or falling land levels.

Fig. 9.34 Models of cuspate forelands: a) Dungeness type, b) Benacre Ness type and c) swash type

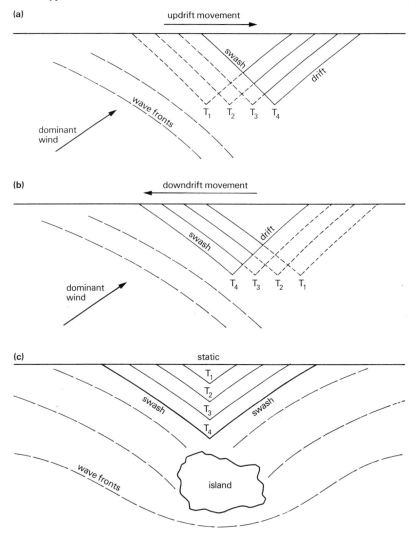

5. Bars

These are accumulations of sediment usually elongated parallel with the coast and submerged during part or all of the tidal cycle. They may be due to a variety of processes but it is assumed that they are the precursors of barriers and are in the process of being driven landwards.

ASSIGNMENTS

1. *Draw a sketch of the Spurn Head spit from Plate 9.8. Label the proximal and distal ends, the point most likely to be breached, the dominant wave direction, mud and sand flats and the orientation of the wave fronts.*

Fig. 9.35 Loe Bar, Cornwall

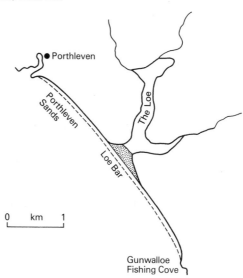

2. *Describe, in a series of numbered statements, the evolution of the Spurn Head spit from 1066 to the present. Divide your statements into a series of cycles (page 263). What is the approximate length of the cycles?*

3. *Describe the conditions necessary to cause cuspate forelands to move: (i) updrift, i.e. against the dominant direction of sediment movement, and (ii) downdrift on a coast.*

4. *Chesil Beach is shown in Plate 9.6. Use an atlas map and the photograph to classify this feature into one of the categories of landform shown in Fig. 9.28.*

E. Changing Sea Levels

In this chapter we have made several references to the rising sea level of the postglacial period. Figure 9.36 shows this rise in more detail. The lowering of temperatures during the Pleistocene glaciation had the effect of transferring huge volumes of water from the oceans to the land thus lowering worldwide sea levels up to 100 m. Such changes are referred to as *eustatic* and affected coasts throughout the world at the same level. This happened not only during the last, Devensian, glaciation but in each of the preceding phases. The curve in Fig. 9.36 shows the rise in sea level since the end of the Devensian glaciation as the melting ice returned water to the oceans.

1. Evidence of former eustatic sea levels

The earlier phases of low and high sea levels have left little sign of their existence on our coasts. The high sea level of the most recent, Ipswichian, interglacial are most clearly preserved in *raised beaches* outside the limit of the Devensian glaciation (page 206). The cross-section of the deposits at Godrevy, St Ives, Fig. 9.37, shows that the sea level was about 4 m above its present level

Fig. 9.36 The rise in eustatic sea level from 20 000 years B.P. to present (After Fairbridge)

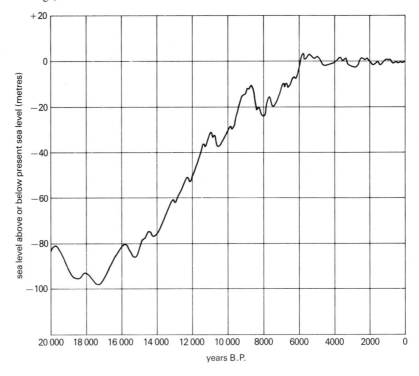

Fig. 9.37 Deposits at Godrevy Point, Cornwall (After West)

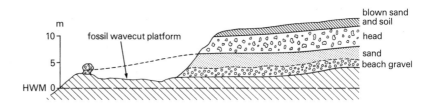

and a platform was cut at this height. A gravel which rests on the platform contains marine shells giving conclusive proof that the sediments represent a fossil beach. The deposits above this were laid down in the Devensian period and are the product of weathering and erosion in a cold climate (page 216).

2. Evidence of lower sea levels

The lower eustatically controlled sea levels during the glacial period enabled rivers to cut down their valleys below the present sea level. On the rise of sea level these valleys were drowned. If the valley remains unfilled and has a typical 'dentritic' pattern of tributaries, it is termed a *ria*. The Loe in Fig. 9.35 must have been a ria before the bar blocked the entrance. The River Fal in Cornwall is a more typical example of a ria form with branches extending to

Truro about 18 km from the open sea. This is slowly being infilled by sediment brought down by the rivers draining from St Austell Moor.

Drowned glacial valleys are termed *fiords* and these in an upland coastal area may be more than 1200 m deep. At the mouth there is usually a shallower area sometimes less than 100 m deep called the *threshold*. Fiords form wide rectangularly branched inlets, in the case of Sogne Fiord in Norway more than 100 km long (Fig. 9.38). The origin of thresholds is difficult to explain and

Fig. 9.38 Long profile of Sogne Fiord, Norway

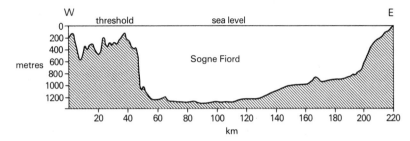

possibly they may have a number of causes. They may mark the point where the glacier began to float and ceased to erode, or they may be large terminal moraines (page 167). A third possibility is that the glacier was thinner near the snout and less erosive. None of these explanations is entirely satisfactory, since some thresholds are of solid rock and others appear to be morainic, at least on the surface.

Drowned valleys in lowland glaciated areas have been given the name *fiards*. They are wider and shallower than fiords and tend to be less branched. The valleys running to the sea on the Oslo Fiord region of Norway and the Baltic coast of Sweden are fiards.

Landforms which have been submerged are a much more frequent occurrence than those formed at higher sea levels. This is a result of the postglacial rise of the sea, referred to in Europe as the *Flandrian transgression*, when the sea level reached its present level, about 5500 years ago. This rise was worldwide and the result of water being returned to the oceans as the ice caps of the Devensian glaciation melted. There are more restricted areas where sea level fell locally in relation to the land. In fact what really happened was that the land rose faster than sea level. This rise may be ascribed to two causes. Firstly in *tectonically active* areas like New Zealand, uplift has occurred as part of the process of mountain building. Beach features have formed and been raised faster than the rise of sea level during the past 10 000 years. Secondly, the ice which formed the glaciers and icecaps in the glacial periods depressed the crustal rocks, forcing them down into the more mobile rocks of the mantle. Removal of the weight of the ice enabled the crust to rise again. This process is called *isostatic adjustment* or *isostasy*. Although the ice sheets had almost disappeared by 8000 B.P. (before present), isostatic adjustment is not yet complete in some areas which were near the centre of the ice sheets and which were, therefore, depressed the furthest. In the Baltic, mooring rings which within living memory were at sea level, are now beyond the reach of boat owners; rates of isostatic recovery of up to 20 mm/a have been recorded there.

Plate 9.9 Taa Fiord, Norway (Author's photograph)

3. Isostatic shorelines

Figure 9.39a shows how much uplift has occurred in Scotland in the postglacial period. The nearer the centre of the isobase pattern the greater the uplift. A beach formed 9000 years ago at A has subsequently undergone 14.0 m of uplift and now forms a raised beach 14.0 m above present sea level. A beach at B formed at the same time as A has undergone 3 m of uplift. Beaches between these places are at intermediate heights so that the one beach formed 9000 years ago is now at different heights wherever it occurs. This height depends on the amount of uplift which has occurred at a particular place. It also follows that there must be places where beaches at the same height are of different ages, having been uplifted at different rates (Z in Fig. 9.39b). For example, a beach 10 000 years old may have been uplifted 30 m, a rate of 3 mm/a, while a beach 5000 years old, as a result of its position closer to the centre of uplift, has also been uplifted 30 m, a rate of 6 mm/a. Many of the isostatic raised beaches of Scotland were originally named from their height, for example the 100 Foot Raised Beach. However, this makes no allowance for the differential uplift of Scotland, and beaches are now named after the time when they were formed, for example the Late Glacial Raised Beach. Many of these beaches have been precisely dated using the carbon isotope C^{14} in shell material deposited in the beach. The 100 Foot Raised Beach at St Andrews has been dated to 13 000 B.P. and it lies about 30 m above present sea level. However, 13 000 years ago

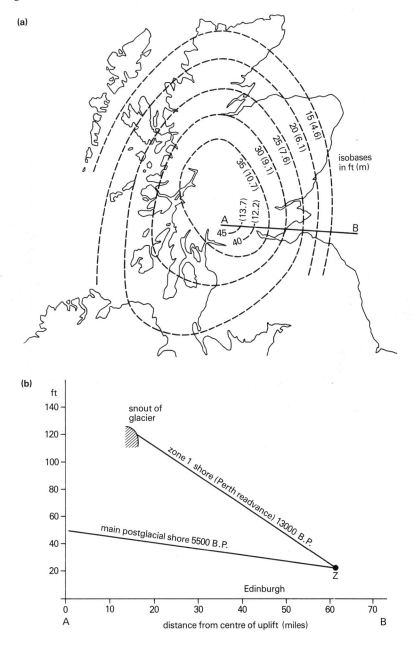

Fig. 9.39 Isostatic beaches in Scotland (After Sissons)

(a)

isobases in ft (m)

15 (4.6)
20 (6.1)
25 (7.6)
30 (9.1)
35 (10.7)
(13.7)
(12.2)

A

45
40

B

(b)

ft

snout of glacier

zone 1 shore (Perth readvance) 13000 B.P.

main postglacial shore 5500 B.P.

Z

Edinburgh

A   distance from centre of uplift (miles)   B

the sea level was 70 m below its present level (Fig. 9.36). In 13 000 years the St Andrews beach must have been uplifted 70 m plus 30 m, a total of 100 m.

On the west coast of Scotland, particularly on exposed coasts like Mull, the raised beaches are marked by rock platforms backed by cliffs (Plate 9.10). In more sheltered areas the beach is cut in glacial deposits and is only marked by a levelling out and smoothing of the depositional topography. On the east coast former sea levels are indicated by deposits of estuarine clays.

Plate 9.10 Raised beach on Mull, Argyll (J.M. Gray)

1. Explain how you could distinguish between a eustatic and an isostatic beach over a long length of coast.

2. What differences would you expect to find between rias and fiords: (i) in longitudinal profile; (ii) cross-section; (iii) in plan? To answer this you may need to refer to Chapters 5 and 6. Illustrate your answers with annotated sketch maps.

3. In Fig. 9.39b the Zone I shoreline (see Chapter 7) has a steeper gradient than the Main Postglacial Shoreline. Explain why this is so.

F. Estuaries and Dunes

1. Estuaries

Most of the larger ports of Britain, for example London, Liverpool, Southampton, Hull and Glasgow, are sited on estuaries. What exactly constitutes an estuary is difficult to define. It is something between an inlet with no infill of sediment, such as a ria or a fiord (page 270), and one with vast amounts of sediment which builds seawards, such as a delta (page 138). Between these two extremes are inlets with varying amounts of infill, all of which loosely come under the heading of estuaries.

There are many inlets on the coast as a result of the submergence of former land surfaces by the Flandrian transgression (page 271). The present environment of these inlets is determined by four main factors: the discharge; the load of rivers entering the inlet; the tidal range of the mouth of the inlet; and its wave environment. In general, little fluvial material escapes from the estuary except where the load and discharge are large. The majority of infill is pushed in from the sea by wave and tidal action. As a result, estuaries are traps for sediment moving along the shore.

(a) *Tidal range*

In macrotidal storm wave environments the characteristic form of the estuary is a triangle, with a river entering at the apex and the base facing the open sea (Fig. 9.40). Within this zone shifting banks of sand, silt or clay are separated by channels. The rising or *flood* tide flows along a different set of channels from those used by the falling or *ebb* tide. At the turn of the tide the ebb and flood channels may have water moving in opposite directions at the same time. The volume of water moving in and out of the estuary in these environments is large and the channels may be very deep as a result of erosion associated with high flow velocities. The constantly shifting sandbanks of such estuaries make them difficult for shipping.

Fig. 9.40 Landforms, sediments and water movements in estuaries

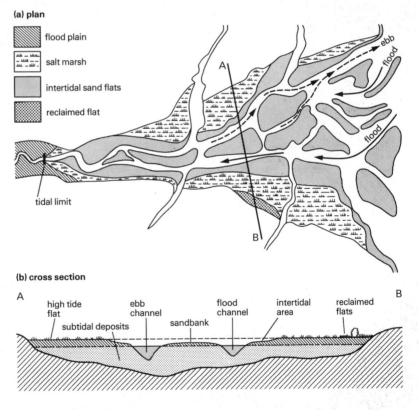

Where the tidal range is smaller, the effect of movement of water in and out of the estuary is less important and wave action tends to produce landforms which block off the estuary from the sea. Where the volume of water contained

in the estuary at high tide is large and the entrance narrow, erosion will maintain a channel to the sea as the water flows in and out along this channel (page 262).

(b) *Sedimentation*

Sedimentation in estuaries is enhanced by their relatively sheltered nature. The flood enters carrying a large volume of fine material in suspension. Where the waves are less active, and at high water when there is little flow, deposition occurs. Because of the Huljstrom effect (page 107) ebb flows cannot remove this sediment from the banks. Fluvial material in suspension is frequently in a charged state and, on meeting the sodium chloride in seawater, particles tend to cluster to form larger particles. This process is called *flocculation*. These particles sink more rapidly than the smaller ones and so speed up the process of deposition.

(c) *The role of vegetation*

In the sheltered wave environments of estuaries vegetation plays a far more important role in deposition than on beaches. The depositional area and the vegetation in the estuary is divisible into three areas (Fig. 9.40). The *subtidal* zone is usually bare sand and because it changes rapidly and is normally submerged little vegetation grows there. The *intertidal* zone is a sloping surface, exposed between the tides, which is colonised by algae and plants like *Salicornia*, the marsh samphire. The *high tide flat* is almost horizontal and is more densely vegetated. The vegetation stabilises the loose surface of the sediment preventing erosion; it also prevents water draining from the area very quickly, restricts erosion and encourages deposition by acting as a trap for sediment. In many British estuaries a prolific cross-bred grass, *Spartina townsendii*, has so

Plate 9.11 Vegetation in estuarine environment (P. Niedzwiedski)

speeded deposition since it was introduced at Southampton in 1875 that many small ports have become isolated from the sea. The Dee estuary and Poole Harbour have large areas of *Spartina*. The build up of sediment eventually leads to the take-over of the salt marsh by more terrestrial species as the salt is leached from the surface by rainwater.

2. Coastal dunes

Dunes are ridges of sand formed by the wind on the landward side of beaches. They form best where there is a wide foreshore which dries out between the tides, a feature usually found on macrotidal coasts. If the beach material is in the sand classes $-1\ \varnothing + 4\ \varnothing$ (page 9), strong onshore winds will dry out

Fig. 9.41 Dune systems and their development (After Ranwell)

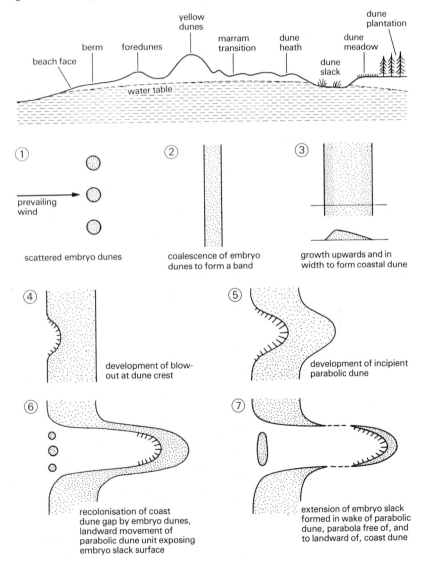

the beach and remove large volumes of sand to build dunes. In tropical areas, where wind speeds are generally low, dunes are less frequent. They are a common occurrence in storm wave environments.

Most coastal dunes in humid areas are halted by vegetation which makes them assume shapes different from desert dunes. *Foredunes* accumulate at the top of the beach where vegetation traps sand. This forms the core of the dune and, as it grows by accretion, species like *marram grass* keep pace with it. When growth stops the marram dies and other grasses take over. Frequently the front edge of the foredunes are cut back by the storm waves so that they often have a gently curving face towards the sea (Plate 9.12). Occasionally, foredunes form long ridges parallel to the beach but, characteristically, they are irregular, with the highest dunes lying nearest the sea.

Erosion of previously formed dunes by the wind causes *blow-outs*. The centre of the blow-out moves rapidly while the limbs are retarded by the vegetation so that an arcuate form, concave seawards results. With a constant and strong wind direction these may develop into parabolic dunes (Fig. 9.41).

The dune system shown in Plate 9.12 is in retreat. Storm waves are cutting into the foredunes, and the pine trees, which may be 60 years old, have died

Plate 9.12 Dune erosion at Formby, Lancashire (Author's photograph)

and toppled over as the water table has been lowered. There is evidence of earlier phases of dune movement. A palaeosol, a buried soil, forms a dark horizon rich in humus across the newly cut cliff. This must have formed when there was little sand movement and a stable vegetation cover, and later inundated as blow-outs and erosion moved sand inland again.

On the Moray Firth in Scotland the Culbin Sands have advanced 5 km inland since a violent storm began the process in 1694. The villages and woods covered by the dunes since this time have subsequently been re-exhumed by the wind.

G. Coastal Classifications

In this chapter we have used several classifications of coasts: Atlantic-Dalmatian, page 243; storm-swell wave, page 235; macro-meso-micro tidal, page 236; submerged-uplifted, page 269; and the chapter itself is divided into erosion and deposition. These classifications are either for convenience of *description* or, more usefully, they differentiate between features on the basis of their origin, that is, they are *genetic*. We have used a combination of tidal and wave environments to give categories like macrotidal storm wave environments and then used these classes to explain the difference between the coast of Britain and other areas of the world (page 275).

If we combine uplift-submergence and erosion-deposition to form a classification, all the world's coasts will fit into the scheme. Valentin produced such a classification which is shown diagrammatically in Fig. 9.42. A coast which is being uplifted and on which deposition is prevalent is the most rapidly advancing type. A coast which is submerging and eroding rapidly retreats. In both these cases the two factors act in conjunction to maximise the effect. Uplift and erosion operate in opposition however, the one tending to cause advance while the other causes retreat. If the two are equal, then the coast is static and does not change its geographic position. In the same way deposition and submergence act in opposition to cause advance and retreat respectively. The origin of the axes in Fig. 9.42 indicates a coast on which there is no geomorphic activity. A particular coast is represented on this diagram by a single point determined by the rate of uplift or submergence and erosion or deposition.

Point A on Fig. 9.42 represents a coast which has been uplifted 100 m, but which has also been subject to rapid erosion and has retreated 1 km. Such a

Fig. 9.42 Valentin's system of coastal classification

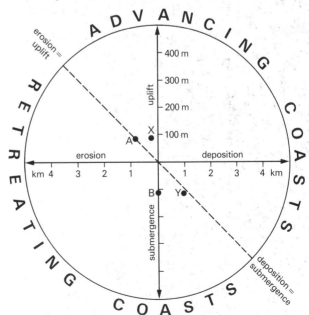

situation might be found at present on a tectonically active coast in a storm wave environment, such as New Zealand or the Pacific coast of Alaska. Point B represents a coast of negligible erosion but which has been submerged 100 m. This could be Cornwall, where the Flandrian rise in sea level has modified the granite cliffs only in detail.

In practice it is not always possible to measure the amount of erosion on a coast and we have to be satisfied with calling it either erosional or depositional. The assemblage of landforms along the coast gives us this information and also a measure of the rise or fall of the relative levels of land and sea. The rias and fiords on page 270 are indicative of a relative rise of sea level since their formation, while raised beaches are evidence of relative falls of sea level. The occurrence of both fiords and raised beaches on the west coast of Scotland illustrates how complex these changes can become.

H. Coastal Protection

The changes at the edge of the land have possibly affected humans more than most geomorphological changes, other than those almost instantaneous catastrophes such as landslides (page 53). Most human effort has been directed into holding the sea at bay. This has been achieved to greatest effect along the coast of Holland. Where ports are located, an enormous effort is expended to keep them open. Both these require greater work on lowland coasts where the effects of coastal changes are more pronounced, a fact reflected in the migration of tanker terminals to upland coasts in Britain.

In coastal areas where beach drifting is a major process, and a hazard, local authorities have erected *groynes* to arrest the sediment movement and form a protective barrier of beach material along the coast. The groynes form a sediment trap in the same way that a seaward change in the orientation of the coast described on page 258 causes a beach to accumulate. The sediment piles up on the updrift side of the groyne until it is high enough to overtop it or it can escape by drifting round the end (Fig. 9.43a). This simple and effective measure can often have disastrous consequences for adjacent authorities on the downdrift side. Deprived of their supply of sediment, their beaches are eroded as the drift moves the beach material away downdrift and wave energy is used to attack the cliff at the back of the beach.

Where wave attack is strong, sea walls have been built to protect the land from erosion. In some cases these are true vertical walls, often with a lip at the top to reflect the force of the wave and prevent sediment being thrown over the wall. Other designs have a long sloping ramp which absorbs the wave energy more gradually (Fig. 9.43b). Large boulders dumped at the top of the beach have a similar effect and act like a shingle beach, allowing the water to percolate through to reduce the effect of the backwash (page 254). Some very large breakwaters have been built using concrete structures too large to be moved by waves but allowing percolation. These break the force of the waves and are cheaper to construct than solid walls.

Exceptional conditions have occurred which have breached sea defences. One of the most disastrous of these was on the night of 30 January 1953 and caused widespread flooding on the coasts of eastern England and Holland. A very deep depression centre travelled south down the North Sea basin (Fig.

Fig. 9.43 Coastal protection structures

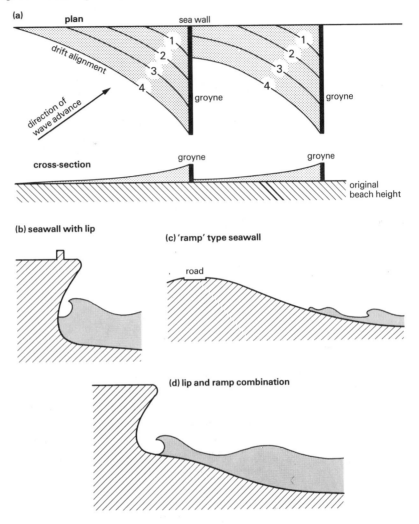

(a)

plan

drift alignment

direction of wave advance

sea wall

1
2
3
4

groyne

1
2
3
4

groyne

cross-section

groyne

groyne

original beach height

(b) seawall with lip

(c) 'ramp' type seawall

road

(d) lip and ramp combination

9.44). The low atmospheric pressure caused the sea level to rise about 0.6 m above that predicted from astronomical data. The winds in the north of the area, estimated to have reached 280 km/h where they were unimpeded by surface obstacles, pushed more water into the basin from the Atlantic. This huge, wave-like tide swept around the shallow basin of the southern North Sea in an anticlockwise direction and, unable to escape through the Straits of Dover, it turned north to travel along the coast of Holland. In estuaries the height of the water was accentuated, the Thames reaching 3 m above the predicted level. At points some distance from the amphidromic points, such as the Yorkshire coast, the level rose by 2.7 m. Such a phenomenon is called a *storm surge*. Its effect was disastrous. Low-lying land along the coasts of England and Holland was flooded by seawater as sea walls were battered by the gigantic waves, and by freshwater as rivers were ponded back by the extraordinarily high tide. The waves cut back cliffs 2 m high south of Lowestoft by 30 m

Fig. 9.44 The North Sea storm surge, 1953 (After King)

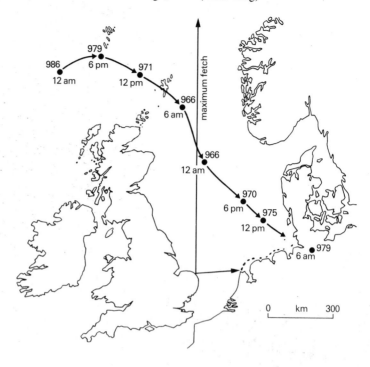

during one high tide. Beaches were lowered by 2 m along the Lincolnshire coast and damage was greatest where the beach was low and narrow. Concrete sea walls were undermined by the destructive action of large storm waves up to 4 m high.

The effect of this surge could have been worse. Although it occurred at the spring tide, two weeks later would have brought it to the highest tide of the year. The peak of the tide and the surge peak were not exactly coincident. Had the surge occurred two hours later the tide would have been even higher.

Dune systems, which are often in the front line of coastal defences, are prone to erosion by both the sea (Plate 9.12) and by the wind (page 281). Foundations for walls and breakwaters are difficult to build on sand and since dunes tend to be of low economic value they often tend to be left unprotected. Wind erosion, in the form of blow-outs (page 277), is a serious problem in areas intensively used for recreation. Paths through the dunes weaken and erode the vegetation mat. The trampled areas are exploited by the wind and become blow-outs, which may enlarge and result in the wholesale movement of the dunes inland. A number of methods are used to combat this destructive advance. Paths can be re-routed so that they lie transverse to the wind and cease to act as 'wind tunnels'. Walkways laid over the vegetation prevent trampling. In extreme cases dunes may be closed to public access.

To stabilise blow-outs, marram grass (page 278) has been planted for centuries, and brushwood mats and fences reduce the wind speed near the ground encouraging sand deposition. At Camber in Sussex a mixture of latex and grass seed was sprayed onto dunes which had been shaped by a bulldozer. This was later made into part of a golf course. Sand quarrying from dunes is more easily controlled by legislation. At Gwithin, near St Ives in Cornwall, and Ainsdale in Lancashire quarrying has been prevented by the local authority.

ASSIGNMENTS

1. *Explain the processes which lead to the infill of estuaries with sediment. Poole Harbour in Fig. 9.31 is an estuary which has a history of changes at its mouth. How do such changes (page 264) influence deposition in the estuary?*
2. *Discuss the relative merits of the methods of preventing coastal erosion on a coast like that of Norfolk. Reference to Ordnance Survey maps and to Fig. 3.17 will give you information on drift and geology. Fig. 9.44 may also be useful.*

Key Ideas

A. *Introduction*
1. Waves transmit *energy* derived from the *wind* across the surface of the sea.
2. At the coast, wave energy is translated via the breaking wave into *work*.
3. The size of the wave and hence the energy is determined by the *wind velocity* and *duration* and the length of *fetch* across the open sea.
4. Tides accomplish little work themselves but spread the effective height range of the waves.
5. The geomorphological environment of any coast may be described with reference to its *wind* and *tide regimes*.

B. *Coastal erosion*
1. The form of the coast is a function of lithology, geological structure and history of the coast. The wave/tidal environment is important in determining the landforms which develop.
2. The configuration of the coast causes wave energy to be greatest on headlands and weakest in bays. A balance develops between resistance to erosion of the headlands and the degree of indentation of the coast.
3. The processes of mass wasting operate on coastal slopes with the sea acting as the *basal removal agent*.
4. *Shore platforms* are widespread and frequently independent of structure, particular types being developed in different wave/tidal environments.

C. *Beaches*
1. *Beaches* are composed of the uncemented products of rock breakdown and respond rapidly to changes in wave configuration.
2. The *shape* and *orientation* of beaches is dependent on *dominant wave direction, supply of beach material* and *shape of the coast.*
3. *Swash aligned beaches* lie parallel to the dominant wave fronts.
4. Beaches which lie at angles to the dominant wave directions indicate lateral sediment transport.

D. *Constructional landforms*
1. Major *spits* occur where there is a change in coastal orientation. They are *drift aligned*.

2. Similar morphological types may evolve in different ways. A study of their history is needed to distinguish between them.
3. Coastal depositional landforms evolve rapidly in geomorphological terms. Significant changes can be observed in historical time.

E. *Changing sea levels*
1. *Eustatic* changes of sea level are worldwide and are due to changes in the volume of water in the oceans or to major changes in the volume of the ocean basins themselves.
2. *Isostatic* changes are due to the depression or uplift of the crust as a response to changes of the weight of the *load on the crust*. *Tectonic* changes are the result of major earth movements and are generally confined to the margins of the plates of the crust.
3. Eustatic sea level changes produce features which are essentially at the same height over a large area.
4. The *postglacial eustatic sea level rise* means that the majority of recent features are submerged.
5. Isostatic changes are unequal and give beaches which slope towards the centre of isostatic uplift.
6. There is a competition between postglacial isostatic and eustatic change.

F. *Estuaries and dunes*
1. In *estuaries* tidal flow becomes dominant in terms of erosion and deposition.
2. The *sheltered* nature of estuaries allows vegetation to play a significant role in enhancing deposition.
3. *Dunes*, though aeolian features, are associated with wide beaches from which sand can be eroded by the wind. Under natural conditions *vegetation stabilises* them.

G. *Coastal classification*
1. Coastal classifications may be *morphological* or *genetic*. They are one of the initial stages in the clarification of our ideas on how coasts evolve.

H. *Coastal protection*
1. The principles of shallow water wave changes and deep water reflection are used in the construction and design of sea walls.
2. *Dune stabilisation* is dependent on the establishment of a full vegetation cover.
3. *Coastal defences* must be constructed to withstand the *exceptional event*. The cost of this must be weighed against the benefits in economic and social terms.

Additional Activities

1. The regular cycles of the tides and persistence of waves produce rapid changes on the coast. The processes involved are relatively easy to examine:
   (a) Measure wave period, height and length using a stop watch and marked or measured points on a pier or breakwater. The deep water length (L) is given by multiplying the square of the period (T) by 1.56, i.e. L =

$1.56T^2$. This provides basic information together with meteorological observations for the explanation of changes on the beach.

(b) Surveys of beach profiles using the methods described in Chapter 3 can be used to examine how effective any particular set of conditions are in producing change on the beach. Surveys may be carried out after one tidal cycle or at regular time intervals or following particular storm periods. Beaches with different exposures and with different types of beach material may respond in different ways to the same conditions. Useful comparisons can be made between equally exposed beaches with different materials or between a sheltered and an exposed beach with similar deposits.

(c) The rate of beach drifting can be measured using brightly painted pebbles. A pile of painted pebbles dumped on the beach at low tide amongst similar sized pebbles can be mapped at the next low tide and the movement measured. If conditions are stormy do not expect to recover many of the pebbles, possibly only 1 or 2 per cent may be found. Marine yacht paint is a very good marker.

(d) The shape and roundness of pebbles may change along the beach in the direction of transport. Methods of measuring shape were described in Chapter 5.

(e) Longer term changes on the coast can be assessed by comparing early editions of Ordnance Survey maps with more recent ones. In addition local libraries often have air photographs taken at different times which will enable you to see coastal changes in more detail. Photographs of estuaries where changes in vegetation like *Spartina* can be mapped are very useful. Early records, such as the maps, are less accurate but often give a good idea of some major changes.

(f) Surveys of shore platforms enable correlations to be made between slope, rock type and fetch. These features change slowly and cannot be treated in the same way as beaches. Material transported from the foot of the cliff towards the low water line may show a progressive rounding and may provide a good initial hypothesis for investigation.

. How are coastal erosion and deposition linked together? Using examples, explain why it is not always possible to explain present day landforms by examining present day processes.

. Compare the ways in which a river erodes with wave erosion. Under what conditions are each of these most effective?

. Figure 9.45 shows Holy Island on the north-east coast of England. Explain the origin of the Goswick Sands sand ridge and the dunes which form the The Snook. Keel Head beach probably has very little material drifting onto it; discuss the factors which lead to this. Compare the processes operating between Bride's Hole and Emmanuel Head to those operating between Wide Open and Old Law. What changes are likely to occur on this coast in the future?

Fig. 9.45 Holy Island

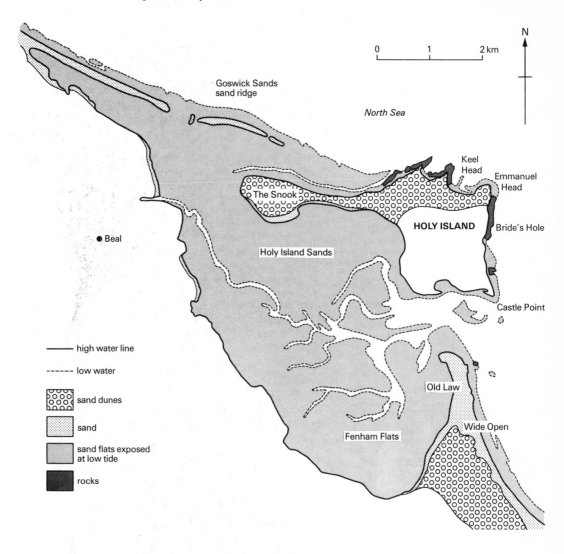

Reading

KING, C.A.M., *Beaches and coasts*, Edward Arnold, 1972

WILLIAMS, W.W., *Coastal changes*, Routledge Kegan and Paul, pages 185–206, 1960

BIRD, E.F.C., *Coasts*, MIT Press, 1968

*DAVIS, J.L., *Geographical variation in coastal development*, Longman, Chapters 3, 5 and 6, 1980

SISSONS, J.B., *Scotland*, Methuen, pages 117–131, 1976

STEERS, *The coastline of England and Wales*, Cambridge University Press, 1964

BLOOM, A.L., *The surface of the earth*, Prentice Hall, Chapter 6, 1969

*PITTY, A., *An introduction to geomorphology*, Methuen, 1971

Index

gt̄
16.9.05

ADOLESCENCE
and EMERGING
ADULTHOOD

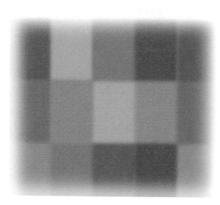

- 2 APR 2006